NIST GCR 02-843-1
(Revision)

Analysis of Needs and Existing Capabilities for Full-Scale Fire Resistance Testing

Jesse Beitel
Nestor Iwankiw
Hughes Associates, Inc.
3610 Commerce Drive, Suite 817
Baltimore, MD 21227-1652

National Institute of Standards and Technology
Technology Administration, U.S. Department of Commerce

Errata

Replace

NIST GCR 02-843 with NIST GCR 02-843-1, (Revision 2008)

The Building and Fire Research Laboratory has made corrections to NIST GCR 02-843, December 2002.

Title: Analysis of Needs and Existing Capabilities for Full-Scale Fire Resistance Testing. October 2008.

Jesse Beitel
Nestor Iwankiw
Hughes Associates, Inc.
3610 Commerce Drive, Suite 817
Baltimore, MD 21227-1652

Corrections:

September 2008: Made changes to Table 2.1, Summary of Multi-Story Building Fires with Collapses (4 or more Stories), page 24 and continuing on page 25.

October 2008: Inserted Note page ii, and made a change to section, 2.4 Conclusions, first paragraph, last sentence, page 22.

NIST GCR 02-843-1
(Revision)

Analysis of Needs and Existing Capabilities for Full-Scale Fire Resistance Testing

Prepared for
U.S. Department of Commerce
Building and Fire Research Laboratory
National Institute of Standards and Technology
Gaithersburg, MD 20899-8660

by
Jesse Beitel
Nestor Iwankiw
Hughes Associates, Inc.
3610 Commerce Drive, Suite 817
Baltimore, Maryland 21227-1652

October 2008

U.S. Department of Commerce
Donald L. Evans, Secretary

Technology Administration
Phillip J. Bond, Under Secretary for Technology

National Institute of Standards and Technology
Arden L. Bement, Jr., Director

Note

This report uses the term "fire-induced collapse" to indicate the failure of a structure, or significant portion of a structure, that could be attributable directly to a fire event in the building. In some cases, the building may have been under construction or in process of renovation, or it may have experienced significant damage prior to the fire caused by a blast, impact, or an earthquake. A subsequent report on the collapse of the WTC towers (NIST NCSTAR 1, September 2005) found that WTC 1 and WTC 2 collapsed due to aircraft impact damage to the structure and fireproofing as well as to fire. Both effects (damage and fire) were equally important. In the absence of structural and insulation damage, a fire substantially similar to or less intense than the fires encountered on September 11, 2001, likely would not have led to the collapse of a WTC tower. On the other hand, it was concluded in NIST NCSTAR 1A (October, 2008) that WTC 7 would have collapsed from fires having the same characteristics as those experienced on September 11, 2001, even without the initial structural damage to the building initiated by the collapse of WTC 1. The Building Performance Study conducted by the Federal Emergency Management Agency (FEMA 403, May 2002) refers to the partial collapse of WTC 5 as "fire-induced." No analysis was conducted to determine the relative roles of the initial structural damage and the subsequent fires that led to the fate of WTC 5.

The number of historical building collapses that are identified in this report as "fire-induced" (a total of 22) includes WTC 1, WTC 2, WTC 5 and WTC 7, consistent with the above definition of the term. This does not imply that WTC 1, WTC 2 or WTC 5 would have collapsed due to a severe, uncontrolled fire had the structure and/or insulation not been severely damaged prior to the outbreak of fire.

EXECUTIVE SUMMARY

This program[1] was conducted for The National Institute For Standards and Technology under Contact Number NA1341-02-W-0686. Hughes Associates, Inc. performed this work with assistance from Greenhorne & O'Mara, Inc. and Thornton-Tomasetti-Cutts, LLC. The study was commissioned to analyze the needs and existing capabilities for full-scale fire resistance testing of structural connections. The Scope of Work consisted of three separate tasks. The tasks were:

Task 1. Identification Of Building Collapse Incidents - The objective of this Task was to conduct a survey of historical information on fire occurrences in multi-story (defined as four or more stories) buildings, which resulted in full or partial structural collapse.

Task 2. Survey Of Fire Resistance Test Facilities - The objective of this Task was to perform a survey of private and public facilities capable of testing the structural integrity of building elements under fire conditions.

Task 3. Needs Assessment - The objective of this Task was to perform an assessment of the need for additional testing and/or experimental facilities to allow the performance of structural assemblies and fire resistance materials to be predicted under extreme fire conditions within actual buildings; and if a need does exist, options for meeting those needs.

In Task 1, the search for this data was conducted using three principal sources: news databases, published literature, and direct inquiries to key individuals and organizations. Even though the task objective was to identify multi-story fire-induced collapses, other useful and pertinent information on major multi-story fires without collapses, but with major structural damage was obtained. The results of the world-wide survey indicated that a total of 22 fire-induced collapses were identified spanning from 1970 to the present. The 2001 World Trade Center (WTC) collapses accounted for four of these events. Seven major multi-story fire events were also identified as having significant structural damage due to a fire, but did not exhibit collapse. While this total number of fire events may appear low (average of one/year) these fire events are high consequence events with respect to economic costs and potential for loss of life and/or injuries.

In Task 2, the survey of the potential testing laboratories was conducted via questionnaires sent to various laboratories, organizations, and individuals. The responses were collated and reported. In general, the existing facilities can perform the standardized fire resistance tests which generally limit the various test parameters with respect to testing individual building elements, size of elements, fire exposure conditions, downward live loading, and measurements obtained. The survey did identify that to some limited extent, some combinations of building elements may be evaluated in the existing test furnaces but no one laboratory can provide all of the combinations of size, capabilities, etc. Also, no facility readily exists that can provide full- or real-scale tests of combinations of building elements or their connections and interactions.

Task 3, the needs assessment found that a more complete understanding of structural performance of building elements when exposed to a fire is required. Structural fire protection

---

[1] HAI Project #5243, Hughes Associates, Inc.

has largely been ignored as a research area in the past 25 years and the WTC collapses have renewed interest in this area of research. It is apparent that research into structural fire protection performance will require unique test facilities to be developed. Specialized test apparatus and instrumentation for large structural assemblies, for real-scale building elements and their connections, and full- or real-scale combinations of elements will need to be constructed. Specialized and enhanced instrumentation will also be needed.

DISCLAIMER

Certain companies and commercial products are identified in this report in order to specify adequately the source of information or of equipment used. Such identification does not imply endorsement or recommendation by the National Institute of Standards and Technology, nor does it imply that this source or equipment or services is the best available for the purpose.

CONTENTS

Page

EXECUTIVE SUMMARY v
1.0 INTRODUCTION 1
2.0 TASK 1: IDENTIFICATION OF BUILDING COLLAPSE INCIDENTS 2
    2.1 Objective 2
    2.2 Survey Scope and Methodology 2
        2.2.1 News Sources 2
        2.2.2 Literature 4
        2.2.3 Direct Inquiries 4
    2.3 Survey Results 5
        2.3.1 Multi-story Buildings With Fire-induced Collapses 5
        2.3.2 Selected High-Rise Building Fires Without Collapses, But With Major Structural Damage 18
        2.3.3 Low-Rise Buildings with Collapse 20
        2.3.4 Fires After Earthquakes 20
    2.4 Conclusions 22
3.0 TASK 2. FIRE RESISTANCE TESTING LABORATORY SURVEY 32
    3.1 Objective 32
    3.2 Survey Scope and Methodology 32
    3.3 Survey Responses 33
    3.4 Discussion of Survey 33
4.0 TASK 3. ASSESSMENT OF NEEDS 45
5.0 REFERENCES 48
APPENDIX A – FIRE RESISTANCE QUESTIONNAIRE 51
APPENDIX B- COMPLETED LABORATORY QUESTIONNAIRES 56

TABLES

Page

Table 2.1   Summary of Multi-Story Building Fires With Collapses 24

Table 2.2   Selected Multi-Story Building Fires With No Collapses 26

Table 2.3   High-Rise Building Fire Experience Selected Property Classes, by Year 1985-98 Structure (Reproduced from Hall, 2001) 27

Table 2.4   Recent Low-Rise Building Fires With Collapses 28

Table 3.1   List of Laboratories/Contacts 38

Table 3.2   Summary of Responses 39

Table 3.3   General Testing Laboratory Information 40

Table 3.4   Vertical Structural Building Elements 42

Table 3.5   Horizontal Structural Elements 43

Table 3.6   Additional Information 44

FIGURES

Page

Figure 2.1a Damaged and Burning WTC 1 (right) and WTC 2 (left) on Sept. 11, 2001 7

Figure 2.1b Fires in WTC 5 7

Figure 2.1c Fire Damage in WTC 5 8

Figure 2.1d Partial Connection and Floor Failures in WTC 5 8

Figure 2.2 Pentagon Collapse from the Sept. 11, 2001 Attack 9

Figure 2.3 Collapse of Burning Apartment Block in St. Petersburg, Russia 10

Figure 2.4 Collapsed Textile Factory in Alexandria, Egypt 10

Figure 2.5 CESP 2 Core Collapse in Sao Paulo 12

Figure 2.6 Katrantzos Department Building in Athens after the 1980 Fire 13

Figure 2.7 Large Lateral Deformations and Failure of Columns at Sixth Floor of Military
  Personnel Records Center 14

Figure 2.8 Boston Vendome Hotel Collapse 15

Figure 2.9 One Meridian Plaza Interior Fire Damage 19

Figure 2.10a Post-earthquake Fires in Kobe, Japan on Jan. 17, 1995 21

Figure 2.10b Burned Area in the Nagata Ward of Kobe 22

Figure 3.1 Fire Exposure Conditions 35

1.0 INTRODUCTION

This study reported here was conducted for the National Institute of Standards and Technology (NIST) under Contact Number NA1341-02-W-0686. Hughes Associates, Inc. performed this work with assistance from Greenhorne & O'Mara, Inc. and Thornton-Tomasetti-Cutts, LLC.

The study was commissioned to analyze the needs and existing capabilities for full-scale fire resistance testing of structural connections.

The Scope of Work consisted of three separate tasks.

Task 1. Identification of Building Collapse Incidents

The objective of this Task was to conduct a survey of historical information on fire occurrences in multi-story (defined as four or more stories) buildings, which resulted in structural collapse. Either partial or total failure of the structural framing, members, and/or connections was considered to have constituted a collapse.

Task 2. Survey of Fire Resistance Test Facilities

The objective of this Task was to survey private and public facilities (for-profit, not-for-profit; academic, local, state and federal government; military and civilian; domestic and international) capable of testing structural integrity of building elements under fire conditions to establish the current global research capabilities in structural fire protection.

Task 3. Needs Assessment

The objective of this Task was to perform an assessment of the need for additional testing and/or experimental facilities to support the development of predictive structural fire protection methods within actual buildings; and if a need does exist, options for meeting those needs.

2.0 TASK 1: IDENTIFICATION OF BUILDING COLLAPSE INCIDENTS

2.1 Objective

The objective of this Task was to survey historical information on fire occurrences in multi-story (defined as four or more stories) buildings, which resulted in structural collapse. Either partial or total failure of the structural framing, members, and/or connections was considered to have met the definition of "collapse."

Information sought included:

- Date and location of fire;
- Type of building, its occupancy use, construction type, number of stories, etc.;
- Cause and extent of the fire event;
- Description of the structural collapse;
- Additional information concerning the event, as available; and
- Reference for the fire event information to include literature citations, etc.

2.2 Survey Scope and Methodology

The historical search for catastrophic multi-story fires included incidents dating back to the 1950's, or earlier, with emphasis on those which occurred in North America. In addition, similar events that occurred throughout the world were also solicited and captured as available. The search for this data was conducted using three principal sources: news databases, published literature, and direct inquiries to key individuals and organizations. Even though the task objective was to identify multi-story fire-induced collapses, other useful and pertinent information on major multi-story fires without collapses was obtained during the normal course of this survey process. For the sake of completeness, there is also a short presentation and discussion of low-rise (less than 4 story buildings) fires with structural collapses.

2.2.1 News Sources

With the volume of information reported electronically concerning the collapse of the World Trade Center (WTC), it was necessary to filter much of the web-based information to find the additional cases. Standard Internet search engines did not yield many specific fires with collapse. Consequently, news sources were consulted for their record of major fire events. LexisNexis™, a powerful full text news database, provided the ability to search for occurrences where both *fire* and *collapse* appeared in the same sentence. The keyword *story* was also included to determine whether the case fit the desired multi-story profile. The WTC Towers were removed from the search in news articles after September 11, 2001. Major world news sources, as well as local US sources, were searched from 1997 to present, the primary time frame included by electronic database sources.

The news sources that were searched included:

LexisNexis™ Academic Universe News Category Online

<u>World News</u> (various news sites outside U.S)
General News>Major Newspapers> 1997 to present

<u>US News</u>(various news sites in U.S.)
Midwest > 1997 to present
Northeast > 1997 to present
Southeast> 1997 to present
Western> 1997 to present

ABC News
www.abcnews.com

NBC News
www.msnbc.com

BBC News
http://news.bbc.co.uk /

NIOSH Firefighter Fatality Investigation Reports
http://www.cdc.gov/niosh/facerpts.html

California Occupational Health Surveillance and Evaluation Program
http://www.dhs.cahwnet.gov/ohb/OHSEP/FACE/97CA003.htm

US Fire Administration
http://www.usfa.FEMA.gov/dhtml/inside-usfa/nfdc-data10.cfm

Interfire.org
http://www.interfire.org/res_file/pdf/Tr-061.pdf

NFPA
www.NFPA.org

The world search was divided into North/South America, Europe, Asia/Pacific, and Middle East/Africa. The US was divided into Midwest, Northeast, Southeast, and West regions.

Commercial news websites did not provide much useful fire data. The one exception was the BBC, from which two major fire-induced collapses were identified. Other news websites did not have powerful search capabilities.

Data sources focused on firefighter and other fatalities were also searched for fire-induced structural failures. The National Institute for Occupational Safety & Health (NIOSH) Firefighter Fatality/Injury Investigation Reports online included some fire-induced structural collapses. The US Fire Administration National Fire Incident Reporting System (NFIRS) database was general in nature and did not indicate whether structural failures had occurred. It was apparent that within the fire community's reporting and tabulation of fire events and fatalities, the occurrence of either partial or total collapse of a structure was not specifically recorded or reported. This seriously limited our ability to understand the nature and extent of fire-induced structural collapses.

2.2.2   Literature

The engineering literature database "Compendex" was searched using the keyword *collapse*, but most documents covered only analytical simulation or modeling work. NIST's FireDoc database similarly did not provide any actual cases.

Many National Fire Protection Association (NFPA) reports and other available publications were reviewed. The most pertinent survey references are given in Section 5.0. There is no single, or central systematic repository of data on the fire-induced structural collapse, either in the US or abroad.

2.2.3   Direct Inquiries

To supplement the broad and extensive news and literature searches, direct contacts were made with individuals and organizations that were expected to have authoritative information on historical fire-induced collapses. The list of domestic and international professional organizations, companies, and/or governmental agencies contacted included:

ABS-EQE
American Concrete Institute (ACI)
American Institute of Steel Construction (AISC)
American Iron and Steel Institute (AISI)
American Society of Civil Engineers (ASCE)
Arbed Steel
British Constructional Steel Association (BCSA)
Canadian Institute of Steel Construction (CISC)
Construction Technology Laboratory (CTL)
Corus-British Steel
CTICM, France
Disaster Prevention Institute, Kyoto University, Japan
Factory Mutual Research Corporation (FMRC)
Institute for Business and Home Safety
International Association For Fire Safety Science (IAFSS)
Isolatek International
Mexican Institute of Steel Construction (IMCA)
National Fire Protection Association (NFPA)

National Institute for Fire and Research (Japan)
National Research Council of Canada (NRCC)
Nucor-Yamato Steel Corp.
Society for Fire Protection Engineers (SFPE)
Victoria University of Technology, Australia

In addition, a survey request for information was sent to several prominent engineers and consulting firms.

Special acknowledgment for their contributions of pertinent papers, reports, and other information in response to these direct requests is given to Dr. Rosario Ono, Institute for Technological Research, Dr. Farid Alfawakhiri, American Institute of Steel Construction (AISC), Mr. John Dowling, Corus-British Steel, Mr. Manny Herrera of Isolatek International, and Mr. Robert Duval and Ms. Teresa Frydryk of NFPA.

2.3   Survey Results

The results of this study for fire-induced collapses of multi-story buildings are given in Section 2.3.1. Section 2.3.1 includes a summary table and further discussion of these important incidents. In Sections 2.3.2 and 2.3.3, ancillary information on fire damage and casualties is also presented for major multi-story fires without collapses, and for low-rise building fires with collapses. Section 2.3.4 discusses fires after major earthquakes and conclusions are given in Section 2.4.

2.3.1   Multi-story Buildings with Fire-induced Collapses

Table 2.1 presents the summary findings of the survey for this category of fire incidents in chronological order, starting with the most recent events (all tables found at the end of each Section). A total of 22 fire-induced collapses were found. Table 2.1 contains the basic highlights of the building fire and further explanations and descriptions of most of these catastrophic events are included in this section.

The events related to the September 11, 2001 terrorist attacks on the World Trade Center (WTC) complex in New York are the dominant fire and collapse events of this survey (represented as 4 separate incidents). The Federal Emergency Management Agency (FEMA) report (FEMA 403) published in May, 2002 is a notable reference that provides an overview of each of the directly affected buildings (WTC 1, WTC 2, WTC 5, WTC 7). The FEMA report also contains the structural, architectural, and fire resistance design characteristics of the various WTC buildings and the nearby damaged structures. Even though the fires were the final destructive force in the collapse of the WTC Towers, their pre-fire condition with extensive structural and fire protection system damage due to the unprecedented impacts by the two large commercial aircraft presented a unique hazard.

The 9-story WTC 5 building suffered extensive structural damage due to WTC 1 debris, which subsequently led to partial collapses of 4 floors due to fires. Interior floor beam splices between supporting column-tree stubs and beams were identified as the point of fire-induced failure of these several floor bays.

Figure 2.1a shows the fires in the damaged WTC 1 and 2 Towers before collapse, while Figure 2.1b shows the blaze in WTC 5. The representative fire damage in WTC 5 is given in Figure 2.1c, while the aforementioned column tree connection and floor failures are pictured in Figure 2.1d.

The full collapse of the 47-story steel-framed WTC 7 has been attributed to fire causes, which occurred approximately eight hours after the collapse of WTC 1. The structural steel framing and fire protection in WTC 7 were quite conventional, when the building was placed into service in 1987. Perimeter steel moment frames, 2-story belt trusses, and an interior braced core at the lower levels provided lateral resistance features. The floors were typical steel beams with composite deck and concrete topping. The nature of the September 11th fires and their destructive structural effects are the subject of an ongoing investigation by National Institute of Standards and Technology (NIST).

On September 11, 2001 the 5-story Pentagon building in Washington, DC, was struck by an aircraft, resulting in extensive damage and fire (see Figure 2.2). The Pentagon was constructed between 1941 and 1943 of hardened, cast-in-place reinforced concrete. News, anecdotal and visual observations indicate that there were some partial structural collapses in the Pentagon due to the ensuing fires approximately 30 minutes after the jet impact. An official investigation of the incident is being prepared but has not yet been released. There is no other publicly available information on the nature and progression of the structural damage in the Pentagon resulting from to the jet crash and subsequent fire.

The remainder of this Section will focus on the less publicized, but still significant, fire-induced collapses of multi-story buildings. Brief overviews of these incidents are provided to supplement the Table 2.1 summary. The information on the following incidents was obtained from news media reports, unless otherwise noted with particular citations or references. Longer narratives are provided for the more significant events.

Figure 2.1a   Damaged and Burning WTC 1 (right) and WTC 2 (left) on Sept. 11, 2001

Figure 2.1b   Fires in WTC 5

Figure 2.1c Fire Damage in WTC 5

Figure 2.1d Partial Connection and Floor Failures in WTC 5

Figure 2.2 Pentagon Collapse from the Sept. 11, 2001 Attack

The most recent incident reported was the destruction of the Santana Row, Building 7 in San Jose, CA on August 19, 2002 (Chui, 2002; Gathright, et.al., 2002). This structure was a 5-story wood frame retail and residential complex that was under construction. With incomplete fire protection systems, the building was much more susceptible to fire spread and damage. Witnesses state that the fire started on the roof, where construction work was underway. The entire complex collapsed due to the fire.

Another recent fire-induced building collapse occurred in St. Petersburg, Russia on June 3, 2002 (BBC News Online, 2002; Ottowa Citizen, 2002). This building was a 9-story concrete apartment block that totally collapsed after about a one hour fire (See Figure 2.3). The news services reported only 1 related casualty, with about 400 other residents safely evacuating the burning building prior to collapse. It was reported that ongoing reconstruction work at this site had accidentally ruptured a gas line, which ignited and fueled this fire.

A severe fire in the 21-story Jackson Street Apartment building in Hamilton, Ontario on February 8, 2002 (Stephan, 2002) caused partial collapse of a floor/ceiling assembly. Fragments of concrete, as well as lath and plaster, were reported to have injured several firefighters, but no fatalities were reported.

On February 27, 2001, the 4-story Faces Nightclub and Memories Lounge Bar building in Motherwell, Lanarkshire, UK experienced a total collapse after burning for approximately 2 hours (The Herald (Glasgow), 2001).

Figure 2.3 Collapse of Burning Apartment Block in St. Petersburg, Russia

A fire-initiated collapse of a 6-story reinforced concrete textile factory occurred in Alexandria, Egypt on July 19, 2000 (Reuters News, 2000; BBC News, 2000). The fire started in the storage room at the ground floor. Fire extinguishers were non-functional, and the fire spread quickly before firefighters arrived. Approximately nine hours after the start of the fire, when the blaze seemingly was under control and subsiding, the building suddenly collapsed, killing 27 people. Figure 2.4 shows a photograph of this collapse.

Figure 2.4 Collapsed Textile Factory in Alexandria, Egypt

Between 1993 and 2000, six other fire-induced collapses were reported in the news media (Table 2.1). The buildings were all four to eight story, mostly residential, buildings in the USA. The "old" wood Vandergrift apartments in Pittsburgh, PA (Belser, 2000) totally collapsed on

May 7 2000. The collapse was initiated by a back wall failure, causing several fatalities. On February 9, 2000 in Newton, MA (Burke, 2000), a 4-story, brick commercial building collapsed about one hour after the start of the fire, with the upper stories collapsing onto the floors below. One casualty resulted, probably due to the fire itself. This building had been undergoing roof, facade, and electrical renovations. The remaining incidents between 1993 and 2000 caused partial structural collapses with no reported fatalities, such as the roof and/or floor collapses in the Effingham Plaza Nursing Home in Portsmouth, VA on April 6, 1998 (Portsmouth Times, 1998), in the Coeur de Royale Condominium in Creve Coeur, MO on August 25, 1994 (Bell, 1994), and in the Central Square Apartments in Cambridge, MA on Oct. 4, 1993 (Tong, 1993). Apartments in Bronx, New York experienced collapse of the entire rear of the building on April 5, 1994 (Onishi, 1994), killing three people.

On May 21, 1987, Sao Paulo experienced one of the biggest fires in Brazil, which precipitated a substantial partial collapse of the central core of the CESP Building 2. This was a 21-story office building, headquarters of the Sao Paulo Power Company (CESP). Buildings 1 and 2 of this office complex were both constructed of reinforced concrete framing, with ribbed slab floors. According to Berto and Tomina (1988), these two buildings had several unique internal features and contents. Both buildings retained their original wood forms used for pouring the concrete floor slabs. Low-height plywood partition walls were also installed in the interiors. The ceiling in CESP 1 was made of plywood attached to the wood forms, while plaster tiles covered the ceiling in Building 2. Both buildings had automatic fire detection and manual fire alarms, but no automatic sprinkler system. Six footbridges connected these 2 buildings to permit convenient pedestrian access at various levels across a separation distance of 9.5 m.

The CESP fire started on the fifth floor of Building 1 from electrical causes, and progressed rapidly upward due to the combustible (wood) ceilings, the underlying formwork, and lack of adequate compartmentalization. Due to the severity of the Building 1 fire and other factors, it not only spread within this building, but also to the companion Building 2. The subsequent fires in CESP 2 ignited simultaneously on several floors due to the high thermal radiation from the original CESP 1 fire. In the presence of similar combustible partition walls and wood floor formwork, the CESP 2 fire also spread quickly. Firefighting efforts in both buildings were unsuccessful. Approximately two hours after the beginning of the fire in CESP 2, its structural core area throughout the full building height collapsed. This collapse was attributed to the thermal expansion of the horizontal concrete T-beams due to fire exposure. This led to the fracture of the vertical framing elements and their connections in the middle of the building, and the subsequent progressive loss of gravity load-carrying capacity (Figure 2.5). After the Building 2 collapse, the fire in Building 1, which was burning out, then re-ignited in floors 1-4 due of the flaming debris pile on the lower levels from Building 2. It was reported that this entire incident, from the original fire ignition in Building 1, spread to Building 2, the collapse of CESP 2 core, and the final re-ignition and burning of the lower floors of Building 1 lasted a total of about seven hours. The fire occurred in the evening and no casualties were reported.

Figure 2.5 CESP 2 Core Collapse in Sao Paulo, Brazil

The Alexis Nihon Plaza fire in Montreal, Canada occurred on October 26, 1986 (Isner, 1986). The building was a 15-story steel-framed office building that was built atop a wide 5-story concrete mall and parking garage. This complex included an adjacent 23-story office building and a 32-story apartment building, all supported by this concrete plaza. The 15-story office tower had conventional steel framing, consisting of steel beam and deck floors and steel columns. The beams and floors had a 3-hour fire rating, while the columns had a 2.5-hour rating. All building elements were insulated with spray-applied mineral fiber. There was no automatic sprinkler system installed in the building. On October 26, 1986 a fire began on the 10th floor, then spread to the 11th and 12th floors, and later to the top floor. Approximately five hours after the fire started, a section of the 11th floor collapsed onto the 10th floor. The fire was declared out the following morning. There were no fatalities or injuries.

Isner stated that the partial collapse involved the failure of the ends of several steel girders supporting the 11th floor, as well as an entire 9.1 m by 12.2 m floor section. Isner reported that these partial collapses resulted from weld fractures of clip angles that connected the floor girders to the columns. It was noted that the steel girders and beams of the collapsed floor section were virtually straight and un-deformed, suggesting that the structural members were not exposed to excessively high fire temperatures or stresses.

As a postscript to this incident, Isner indicated that a smaller second fire occurred in this same building almost one year later during its reconstruction. On November 16, 1987 a fire occurred on the 9th floor, which had received little damage during the initial fire. Firefighters successfully extinguished this fire in approximately 45 minutes. Similar to the previous 1986 fire, several floor girders and beams in the area exposed to the fire collapsed locally, due to weld fractures in the clip angle connections between the beams and columns. The steel members themselves were again observed to be straight and undistorted.

Papaioannou (1986) documented two large department store fires in Athens, Greece in 1980. These suspected arson fires occurred on December 19, 1980. The Katrantzos Sport Department Store was an 8-story reinforced concrete building. Its fire started at the 7th floor and rapidly spread throughout the building, due to lack of vertical or horizontal compartmentalization and the absence of an automatic sprinkler system. Collected evidence indicated that the fire temperatures reached 1000 °C over the 2-3 hour fire duration, and the firefighters concentrated on limiting fire spread to the adjacent buildings. Upon extinguishment of these fires, it was discovered that a major part of the 5th through 8th floors had collapsed. Various other floor and column failures throughout the Katrantzos Building were also observed, (see Figure 2.6). The cause of these failures was considered to be restraint of the differential thermal expansion of the structure that overloaded its specific elements or connections.

Figure 2.6 Katrantzos Department Building in Athens after the 1980 Fire

A partial roof and column collapse of the Military Personnel Record Center occurred on July 12, 1973 (Sharry, et.al., 1974). This was a large 86 m by 222 m, 6-story office building constructed of reinforced concrete. The building was located in Overland, MO and was built in the late 1950's. Sprinklers were present only on the first and second floors. The fire was reported to have started on the 6th floor. Due to the high fuel load of 21.7 million record files stored on the 6th floor, the fire burned out of control for 20 hours. The fire was finally extinguished after 4 days. The roof collapse began after approximately 12 hours of fire exposure, and involved 30% of the roof slab above the estimated point of fire origin. Subsequently, most of the remaining freestanding columns on the 6th floor collapsed. Minimal fire damage was experienced below the 6th floor. The collapse and damage were later attributed to the large horizontal expansion of the 18 cm thick, conventional concrete roof slab that was supported by 41 cm reinforced square-tied columns. There were no expansion joints in the floors or roof. Lateral roof displacements of almost 60 cm occurred in one corner. Figure 2.7 shows the extent of the sixth floor horizontal deformation and damage to the concrete columns due to thermal expansion of the roof. The damage to the columns appeared to be analogous to the brittle column failures that have often occurred during earthquakes.

Figure 2.7 Large Lateral Deformations and Failure of Columns at Sixth Floor of Military Personnel Records Center

Another fire collapse occurred at the historic Hotel Vendome in Boston, MA. Built in the late 19th century of masonry and cast iron, this 5-story building was being renovated when a fire started at on June 17, 1972. After burning for almost 3 hours, all five floors in a 12 m by 13.7 m section collapsed (See Figure 2.8). This incident and the subsequent investigation were described in the *NFPA Fire Journal*,(January, 1973). Ongoing renovation work had caused excessive stresses on the bearing wall under a cast iron column, which was apparently already on the verge of overloading before the fire event.

The One New York Plaza fire occurred on August 5, 1970 (N.Y. Board of Fire Underwriters, 1970) in a 50-story office building constructed of structural steel floor and column framing with a reinforced concrete core. The steel beams and columns were fire protected. The fire started on the 33rd floor and lasted for about 5 hours. The resulting damage was restricted to the 33rd and 34th floors, wherein beams were visibly twisted and deflected. More significantly, filler beams in several sections of these floors dropped onto their supporting girders due to the fracture of the end connection bolts. These local connection failures were attributed to isolated unprotected areas of the steel connections, either due to localized removal of the spray-applied steel fireproofing materials to accommodate certain construction details, accidental

Figure 2.8 Boston Vendome Hotel Collapse

localized fireproofing application omissions, or material fall-off. There were no further collapses of the structural framing during this fire. These localized steel connection failures which occurred in the fire at the One New York Plaza were somewhat similar to those that occurred in the 1986 Alexis Nihon Plaza fire and in the 2001 WTC 5 fire.

In order to have been included in this incident tabulation, fire needed to have been judged the proximate cause for the building collapse (partial or total). Hence, any collapses due primarily to explosions, impacts, earthquakes, wind, and other construction or design factors were beyond the scope of this survey, even if fires had developed during the course of these events.

This survey was restricted only to building fires, so fires in such non-building structures as tunnels, bridges, transmission towers, storage tanks, etc. were excluded. Given these restrictions, the major damage and apparently incipient collapse conditions due to the August 27, 2000 fire in the Ostankino Tower in Moscow, Russia was omitted. This 539 m high Tower was Europe's tallest structure used for broadcasting and communications, but technically not a building. It endured a 24 hour fire with four casualties. According to the news sources, much of the prestressed concrete Tower's steel tendons were compromised. Because of this structural damage, some have claimed that the Tower was on the verge of collapse during and immediately after the fire.

Also, the aggregated building destruction and damage in Kuwait due to the 1990 Iraqi occupation documented by Al-Mutairi and Al-Shaleh (1997) on was not included in this survey. This paper was vague with respect to cause and effect documentation, and lacked specific data on any individual buildings. Similarly, the recent incident on July 12, 2002 of a structural collapse and fire of the 10-story storage facility owned by Quad/Graphics Inc. in Fond du Lac, WI, USA was omitted from the survey results. There are major open questions whether a rack system structural failure precipitated the subsequent collapse and fire, or vice versa. There is a private engineering investigation of this event being conducted at this time.

It was equally difficult to evaluate the separate effects of the many fires in buildings that are known to have occurred after major earthquakes. This subject is addressed later in Section 2.3.4.

In summary, a total of 22 cases from 1970-2002 are presented in Table 2.1, with 15 from the US and two from Canada. The number of fire-induced collapse events can be categorized by building construction material as follows:

- Concrete: 7
- Structural steel: 6
- Brick/masonry: 5
- Unknown: 2
- Wood: 2

Three of these events were from the 1970's, three were from the 1980's, four were from the 1990's, and twelve occurred in 2000 and beyond. This temporal distribution was skewed towards more recent occurrences both due to the magnitude of the WTC collapses (4 collapse events) and the enhanced availability of computerized news media data.

The collapse distribution by building story height was as follows:

- 4-8 stories   13
- 9-20   3
- 21 or more   6

Almost 60% of the cases occurred in the 4-8 story building height range, with the remainder affecting much taller buildings. Six collapses occurred in buildings over 20 stories, with three of these occurring at the World Trace Center complex (WTC 1, WTC 2 and WTC 7).

At least four of these fire-induced collapses occurred during construction or renovations, when the usual architectural, structural and fire protection functions were incomplete or temporarily disrupted. Partial collapses (14 events) were the most frequent occurrences, and the three World Trade Center complete collapses dominated the full collapse event total of eight cases. Office and residential were the primary occupancy types in these 20 buildings, as would be expected in multi-story construction. The occupancy distribution is as follows:

- Office: 9
- Residential: 8
- Commercial: 3
- Combined commercial/residential: 2

Among the general observations from this survey of fire-induced collapses of multi-story buildings was that while they are relatively few in number, the consequences were significant, and could have been even worse in terms of human fatalities and economic losses. The fire risk appeared to be slightly higher during building construction and renovation work. Of the 17 fire incidents in the US and Canada, only the Santana Row development collapse in San Jose, CA, occurred outside the northeastern quadrant of North American (North and East of Missouri).

This data demonstrated that buildings of all types of construction and occupancies, in North America, and abroad, are susceptible to fire-induced collapse, particularly older buildings. The annual fire occurrences in the US, according to Hall, (2001), exceeded 10,000 in buildings that were 7-stories or taller. Those that were undergoing repairs or renovations appeared to further increase the fire and collapse risk. If the fire could not be quickly contained and suppressed by sprinklers, firefighters, or other fire protection measures, it posed a serious life safety hazard for any of the building occupants present. Continued fire spread can lead to a partial or total collapse in a multi-story building, compounding occupant losses, as in some of the cases described above.

Difficulties were encountered during this survey in readily identifying news, and other credible sources of historical and technical information on the fire-induced collapses of buildings. The potential data sources were fragmented, often incomplete, and sometimes conflicting. This lack of data and information significantly hampered the development of a more complete understanding of the magnitude and nature of fire-induced collapse. A centralized reliable body of catalogued information on fire-induced building collapses is needed.

The building code and design objectives are to provide sufficient warning and egress time during a fire emergency that would enable the building occupants to safely evacuate, even if there was an eventual structural collapse. Just as for other natural hazards (wind or earthquake), the time, location, and characteristics of the fire are critical in determining the human and property losses. The total deaths reported for the events in Table 2.1 were over 3,000. Over 2,800 occurred in the recent 2001 collapses of WTC 1 and WTC 2.

A fire-induced collapse in a multi-story building can be classified as a low frequency, high-consequence event. Modern society draws much attention to these and attempts to prevent them, much as it does for earthquakes and windstorms. Given that there can be no guarantee that a fire will not occur in a given building, or that it will be successfully contained and suppressed, the fire resistance of the building structure must be duly assessed in its design in order to avoid local and progressive collapses. Since several of these documented cases demonstrated various member and structural connection failures, a better understanding of the response of various building connections to fire is needed. The effects of elevated temperatures on the strength of connectors themselves and on their ductility, as well as how thermal expansion of adjacent heated members affects the stress redistribution in a floor and framing sub-assemblage through its connections, are important issues yet to be resolved. Connections are generally recognized as the critical link in the collapse vulnerability of all structural framing systems, whether or not fire is involved.

2.3.2 Selected High-Rise Building Fires Without Collapses, But With Major Structural Damage

To complement the fire-induced collapse cases described previously, a summary of selected major recent fires in high-rises that did not suffer collapse, but did incur significant structural fire damage, are presented in Table 2.2. The significance of the selected 7 major fire events in Table 2.2 was that even though there was no associated structural collapse, there was significant fire damage and enormous property loss.

Hall tabulated several hundred high-rise structural fires from 1911-present with fatalities in his 2001 NFPA report, using the NFPA high-rise definition of a building of 7 stories or more. Table 2.3 (Hall, 2001) reported all high-rise fire occurrences in the US (by year) from 1985-1998 for four occupancy (property) classes: apartments, hotels and motels, hospitals and care facilities, and offices. The annual fire occurrences in such high-rises ranged from 10,000 to 17,200 per year, with annual civilian deaths between 23 and 110, annual civilian injuries between 554 to 950, and annual direct property damage between $24.9 million to $150.1 million. It would appear that based on the direct property damage estimates, the majority of these fires were small in nature. To include the fires from other high-rise property classes and in residences with unreported heights, a multiplication factor of approximately 33% was suggested by Hall. This increased the annual range of actual high-rise fire occurrences in the US from 10,000-17,200 to 13,330-22,900. If one were to further adjust this historical fire data for the difference in number of stories between NFPA's high-rise statistics and the NIST multi-story definition as being 4 stories or more (i.e., add 4-6 story buildings to the NFPA summary of 7-story and higher), the number of incidents would be higher.

The 1980 MGM Grand Hotel fire in Las Vegas (Clark County Fire Department, 1981) killed 84 people, injured another 679 people, and caused hundreds of millions of dollars worth of property damage (Clark County Report). The First Interstate Bank (Klem, 1988) (4 floors burned out) and One Meridian Plaza (Klem, 1991)(9 floors burned out) in the US, and the Mercantile Credit Insurance Building and Broadgate fires in the UK (Newman, et al., 2000) are notable examples of excellent overall structural integrity under adverse fire conditions. Some casualties and major economic losses were still incurred in these steel-framed buildings. Complete burnouts of several floors destroyed the interior contents and caused substantial and permanent floor sagging and steel beam distortions, as would be expected after a long, severe fire exposure. In the One Meridian Plaza fire, deflection of main support beams were recorded as large as 46 cm, and one entire area of the 22nd floor had deformed by 1.2–1.5 m. (See Figure 2.9). All the buildings listed in Table 2.2, with the exception of One Meridian Plaza, were repaired and returned to service. After extensive investigations and studies, One Meridian Plaza was dismantled for economic reasons.

Figure 2.9 One Meridian Plaza Interior Fire Damage

Sao Paulo, Brazil had two major high-rise fires in the 1970s. The 31-story Andraus building fire on Feb. 21, 1972 resulted in 16 casualties, while the 25-story Joelma fire caused 189 deaths on Feb. 1, 1974. (Hall, 2001 and Willey, 1972.) Both of these office buildings were constructed of reinforced concrete framing, and contained much of the same nonstructural wood combustibles (floor forms, ceiling tiles, and floor covering) as was previously discussed for the 1987 CESP Building collapse in Sao Paulo. The fires caused severe spalling of large portions of the exterior concrete walls, joists, and columns, exposing the reinforcing steel, due to the severe fire temperatures. Nevertheless, both the Andraus and Joelma buildings remained standing without any collapses. The building were subsequently repaired and returned to service.

Two major fire test programs have been conducted in the UK on full-scale, multi-story and multi-bay structures exposed to fires, at the Building Research Establishment (BRE) Laboratories in Cardington. The first series of tests was conducted on a representative 8-story composite steel-framed office building in September, 1996, (Newman, et. al, 2000). Significant fire damage occurred, as expected, but there were no progressive failures, even with unprotected steel floors. Many other technical observations on the overall high temperature behavior of entire steel structures, as opposed to just isolated individual elements of the structure, were also made. During the summer of 2001, BRE performed another major fire test of a 7-story concrete building (Bailey, 2001). Extensive spalling of the floors and significant lateral displacements of the external columns were experienced, but again without failures, as the compressive slab membrane action provided a secondary strength mechanism.

Both of these research programs strongly suggest that further work on the structural fire response of the entire building framing should be conducted to develop a better understanding of structural fire safety, both in steel and concrete construction.

2.3.3   Low-Rise Buildings with Collapse

Fifty-nine low-rise (2-3 stories in height) building fires with structural collapse, shown in Table 2.4, were found using a world wide newspaper search dating back five years. This number of entries were assumed to be only a subset of the total number of actual occurrences in this category based upon the limitations of the time frame of the search. As indicated low-rise, multi-story incidents (2-3 stories) with some type of collapse occur frequently. Because of their smaller size and occupancies, they had relatively less catastrophic outcomes per incident.

Of the 59 low-rise fires identified in Table 2.4, 49 occurred in North America (45 in the US and 4 in Canada). The vast majority of the low-rise fire collapses were apartments and residential dwellings constructed of wood or brick construction. Generally, three modes of partial collapse were present: floor, wall, and roof. There were only 5-6 cases (or about 10% of these 59 incidents) that led to a total collapse of the structure. Floor and roof collapse were the most common types of partial failures. The information found in the news report was generally provided by fire department personnel. Casualty data ranged from injuries to multiple deaths. Some collapses caused the residents and firefighters to become remotely trapped, or instantly buried. This survey focused on merely identifying these representative collapse incidents, considering the large number of records reviewed and others that were available for low-rise incidents.

2.3.4   Fires after Earthquakes

A fire by itself presents a serious emergency condition, but it can be further exacerbated by another hazardous event(s), such as an earthquake, war, explosions or impact. It is well recognized that a cluster of building fires occurs immediately after most major earthquakes in urban areas. The post-earthquake fire losses can sometimes be comparable to those created by the ground movement alone. Under these circumstances, there is an increased risk of fire sources, reduced or strained firefighting capabilities, the presence of structural damage, and damage to the buildings' fire protection systems.

The 1906 San Francisco earthquake was the signature destructive event of the early 20th century that first emphasized the urgent need for additional seismic structural design criteria for buildings in the US. Likewise, the 1906 earthquake brought to light the danger of the many large fires in the city that added numerous, undocumented, structural failures. Over 3000 individuals died, and the earthquake and fires destroyed large areas of San Francisco. Post-earthquake fire danger has long been acknowledged and has often been repeated.

For this reason, the following general data is provided on the fire effects after the more recent strong earthquakes (Sugahara, 1997; NISTIR 6030, 1996; EQE website):

- October 17, 1989 Loma Prieta Earthquake

    San Francisco experienced 22 structural fires and over 500 reported incidents during the seven hours after the earthquake began. It appeared that all of these affected low-rise buildings (with less than 4 stories), and mostly residential buildings.

- January 17, 1994 Northridge, CA Earthquake

  The Los Angeles Fire Department reported 476 fire incidents during almost a 20-hour period following the earthquake, in contrast to a normal daily count of 50-100 incidents. Other surrounding jurisdictions also reported large numbers of post-earthquake fire calls on that day, including 300 from the Ventura County Fire District, of which 20 were reported to be structural fires, 16 from Santa Monica, and one from Burbank.

- January 17, 1995 Kobe Earthquake

  Approximately 100 fires started within minutes of the quake, primarily in the densely populated, low-rise residential areas of Kobe, Japan. It was reported that several large conflagrations had developed within 1 to 2 hours, with a total of 142 fires and numerous collapses and destruction of mostly low-rise, residential/commercial buildings of simple wood construction. Figure 2.10a and 2.10b shows pictures of the many fires in Kobe and their vast destruction after the 1995 earthquake (EQE Website).

Figure 2.10a Post-earthquake Fires in Kobe, Japan on Jan. 17, 1995

Figure 2.10b Burned Area in the Nagata Ward of Kobe

Available historical records do not clearly indicate any specific multi-story building failures due to post-earthquake fires. This may be the result of a lack of disaster data and precise accounting for the resultant building damage or collapses between the earthquake and fire causes. Nevertheless, these types of fires pose a severe risk to all types of construction in their potentially damaged post-earthquake state, both in terms of their reduced structural and fire resistance. While past experience does not provide any direct evidence of such occurrences in multi-story buildings, the possibility for this combined extreme hazard from both earthquake and fire exposures does exist.

2.4 Conclusions

Fire-induced collapse poses a serious risk in low-rise and multi-story buildings, in terms of life safety and property damage. The Great Chicago Fire of 1871 first exposed the enormity and severity of the fire problem on the US. As demonstrated in this survey of the more modern era, the fire-induced collapse hazard is present in all construction types and construction materials, and in various occupancies. The risks associated with a fire in a high-rise building are greater, though, because of the increased population typically associated with high-rise buildings and the increased difficulty in carrying out a full-building evacuation.

It is well known that the largest number of fires occurs in low-rise and residential construction (Hall, 2001). However, more than 10,000 fires per year in the US have been historically reported in high-rise buildings. While this survey confirms that relatively few of these fires caused subsequent structural collapses in multi-story buildings , the consequences of those fire-induced collapses can be enormous. Fire damage to structural members in multi-story buildings can result in large deflections an order of magnitude greater than the elastic defections normally contemplated for serviceability design.

A central repository of well-documented information on building collapses would be highly desirable in order to more readily enable understand the nature and extent of fire-induced collapses, and further engineering and construction advancements for the enhanced protection of public safety.

Proper fire protection design in compliance with the current, building codes is one obvious necessity to help preserve life safety. Besides adequate provisions for automatic suppression systems, egress, compartmentalization, etc., the fire resistance of the structural frame itself must be adequately provided. In order to minimize the possibility of fire-induced structural collapse of significant proportions, additional structural and fire engineering advancements may be warranted for use in multi-story buildings, and other critical facilities, where such collapses would cause unacceptably large consequences. In particular, as emphasized in FEMA 403 and reflected in this survey, more information should be developed with respect to the structural fire performance of various structural framing and connections as part of entire systems in order to assure overall integrity at elevated temperatures for the sake of enhanced public safety.

Table 2.1 Summary of Multi-Story Building Fires With Collapses (4 or more stories)[§]

Building Name	Location	Type of Construction, Material, and Fire Resistance	# Of Floors and Occupancy	Date, Approximate Time of Collapse, and References	Nature and Extent of Collapse (Partial or Total)
Santana Row, Bldgs. 7	San Jose, CA, USA	Wood frame, still under construction, fire protection and sprinklers not completed/functional	5 Commercial/residential	August 19, 2002 Chui; Gathright	Total collapse and destruction
Apartment block	St. Petersburg, Russia	Concrete	19 Residential	June 3, 2002, starting at 1 hour fire duration BBC News Online	Total
Jackson Street Apartments	Hamilton, Ontario Canada	Concrete	21 Residential	February 8, 2002, News	Partial collapse of concrete floor-ceilings
WTC 7	New York, NY, USA	Steel moment frame with composite steel beam and deck floors; fire resistive with sprinklers	47 Office	Sept. 11, 2001 FEMA 403	Total
WTC 5	New York, NY, USA	Steel moment frame with composite steel beam and deck floors; fire resistive with sprinklers	9 Office	Sept. 11, 2001, unknown time, fire burned uncontrolled for more than 8 hours FEMA 403	Partial collapse of 4 stories and 2 bays
Pentagon	Washington, DC, USA	Reinforced Concrete	5 Office	Sept. 11, 2001, 30 minutes after jet impact Official report release pending	Partial collapses of floors and members
Faces Nightclub and Memories Lounge Bar	Motherwell, Lanarkshire UK	Unknown	4 Commercial/residential	February 27, 2001, after 2 hours News	Total
Textile Factory	Alexandria, Egypt	Reinforced Concrete. no sprinklers	6 Commercial	July 21, 2000, after 9 hours of fire Reuters News	Total
Apartment in Vandergrift	Pittsburgh, PA, USA	Wood	6 Residential	May 7, 2000, few hours after fire started News	Back wall fell, initiating progressive collapse
Commercial complex (near Chestnut Hill Mall)	Newton, MA, USA	Brick/masonry	4 Commercial	February 9, 2000, after slightly more than a 1 hour fire News	Collapse started at upper story and progressed
Effingham Plaza Nursing Home	Portsmouth, VA, USA	Unknown	Multi-story Residential	April 6, 1998, fire started on top floor News	Roof collapsed in places

Table 2.1 Summary of Multi-Story Building Fires With Collapses (Cont.) (4 or more stories)[§]

Building Name	Location	Type of Construction, Material, and Fire Resistance	# Of Floors and Occupancy	Date, Approximate Time of Collapse, and References	Nature and Extent of Collapse (Partial or Total)
Coeur de Royale Condominium I-270 and Olive Blvd.	Creve Coeur, MO, USA	Unknown	4 Residential	August 25, 1994 News	Partial collapses of roofs
Apartments, Brooke Ave and 138th St.	Bronx, NY, USA	Brick	5 Residential	April 5, 1994 News	Rear of the building collapsed.
Central Square Apt. Massachusetts Ave. and Douglas St.	Cambridge, MA, USA	Brick	8 Residential	October 1, 1993 News	Collapse of several floors
CESP, Sede 2	Sao Paulo, Brazil	Reinforced concrete frame, with ribbed slabs; no sprinklers	21 Office	May 21, 1987, after 2 hour fire Berto and Tomina	Partial, full height interior core collapse
Alexis Nihon Plaza	Montreal, Canada	Steel frame with composite steel beam and deck floors; fire resistive without sprinklers	15 Office	Oct. 26, 1986, after 5 hour fire, which then continued for 13 hours Isner, NFPA Fire Investigation Report	Partial 11th floor collapse
Katrantzos Sport Department Store	Athens, Greece	Reinforced concrete	8 Commercial	Dec. 19, 1980 Papaioannou	Partial collapses of 5-8th floor, together with various other members, during a 2-3 hour fire
Military Personnel Record Center	Overland, MO, USA	Reinforced concrete, without expansion joints, no sprinklers above 2nd floor	6 Office	July 12, 1973 1974 Fire Journal	Roof and supporting columns partially collapsed 12 hours after fire began
Hotel Vendome	Boston, MA, USA	Masonry with cast iron	5-6 Residential	June 17, 1972, after almost a 3 hour fire News	All five floors of a 40 by 45 ft section collapsed
One New York Plaza	New York, NY, USA	Steel framing with reinforced concrete core, fire resistive with no sprinklers.	50 Office	August 5, 1970 Abrams	Connection bolts sheared during fire, causing several steel filler beams on the 33-34th floors to fall and rest on the bottom flanges of their supporting girders.

§ WTC 1 and WTC 2 were deleted from this table on August 25, 2008. The two WTC towers collapsed due to aircraft impact damage to the structure and fireproofing as well as to fire. Both effects (damage and fire) were equally important. In the absence of structural and insulation damage, a fire substantially similar to or less intense than the fires encountered on September 11, 2001, likely would not have led to the collapse of a WTC tower. (NIST NCSTAR 1, September 2005)

Table 2.2 Selected Multi-Story Building Fires With No Collapses (4 or more stories)

Building Name	Location	Type of Construction, Material, and Fire Resistance	# Of Floors and Occupancy	Date of Fire Incident, and References	Nature and Extent of Fire
One Meridian Plaza	Philadelphia, PA, USA	Steel frame with composite steel beam and deck floors; fire resistive, but sprinklers not operational (retrofit in process)	38 Office	Feb. 23-24, 1991 Klem, 1991	Started Saturday and burned for a total of 18 hours, causing significant structural damage to 9 floors
Mercantile Credit Insurance Building	Churchill Plaza, Basingstoke, UK	Steel frame with composite floor beams; fire resistive, but no sprinklers	12 Office	1991 Newman, et al., 200	Fire burnout of 8th to 10th floors
Broadgate Phase 8	London, UK	Steel composite trusses and beams; mostly not fire protected and without sprinklers	14 Office	1990 Newman, et al., 200	During construction, 4.5 hour fire duration and temperatures reached 1000 °C
First Interstate Bank	Los Angeles, CA, USA	Steel frame with composite steel beam and deck floors; fire resistive; sprinklers not operational	62 Office	May 4, 1988 Klem, 1988	Lasted for about 3.5 hours, causing major damage to four floors
MGM Grand Hotel	Las Vegas, Nevada, USA	Mixed, no sprinklers	26 Resort and casino	Nov. 21, 1980 Misc. News & Clark County Report	Burned for hours
Andraus Building	Sao Paulo, Brazil	Reinforced concrete	31 Office	Feb. 24, 1972 Hall, 2001	Spalling of exterior walls, joists, and columns, exposing reinforcing.
Joelma Building (Crefisul Bank)	Sao Paulo, Brazil	Reinforced Concrete	25 Office	Feb. 1, 1974 Hall, 2001	Spalling of exterior walls

Table 2.3 High-Rise Building Fire Experience Selected Property Classes, by Year 1985-98 Structure (Reproduced from Hall, 2001)

Table 1. High-Rise Building Fire Experience Selected Property Classes, by Year 1985-98 Structure Fires (Continued)

E. Four Property Classes Combined

Year	Fires	Civilian Deaths*	Civilian Injuries	Direct Property Damage (in Millions)
1985	17,200	66*	665	$24.9
1986	15,000	37*	554	$41.5
1987	13,000	55	635	$36.2
1988	14,600	93	778	$102.3*
1989	14,800	110	798	$58.1*
1990	13,300	83	625	$48.3
1991	13,100	23	747	$150.1*
1992	13,600	34*	827	$75.4
1993	12,400	43	701*	$60.8*
1994	11,300	51	950	$56.9
1995	10,000	55*	688	$44.5
1996	12,100	64*	790	$69.1
1997	11,400	33	560	$43.4
1998	10,000	37	680	$41.1

* In 1985, 1986, 1989, 1991, and 1992, there were 24 total office building fire deaths, all in buildings with unreported height. Since high-rise buildings account for about one-eighth of all office fires, it was estimated that high-rise office buildings had three deaths, allocated as one of the eight 1985 deaths; one of the eight 1986 deaths; and one of the eight 1989, 1991 and 1992 deaths to high-rise buildings, choosing 1992 for the latter as four of the eight deaths occurred in 1992. In 1995 and 1996, deaths in facilities that care for the sick were allocated based on the high-rise share of fires. Property damage figures for apartments in 1991 are inflated by problems in handling the Oakland wildfire in the estimates. Property damage figures for office buildings are underestimated due to problems in handling some large-loss fires, such as a $50 million California fire in 1988, a $50 million Pennsylvania fire in 1989, the $325 million One Meridian Plaza fire in Pennsylvania in 1991, and the $230 million World Trade Center incident in 1993, whose more than 1,000 injuries also are not properly reflected in national estimates.

Note: These are fires reported to U.S. municipal fire departments and so exclude fires reported only to Federal or state agencies or industrial fire brigades. Fires are rounded to the nearest hundred, civilian deaths and injuries are rounded to the nearest one and direct property damage is rounded to the nearest hundred thousand dollars. Property damage has not been adjusted for inflation.

Source: National estimates based on NFIRS and NFPA survey.

Table 2.4 Recent Low-Rise Building Fires With Collapses

Building Name or Occupancy	Location	Construction	Stories	Date	Nature and Extent of Fire
Daycare/Factory	Dromore, Ireland	Unknown	3	May 30, 2002	Fire in a former factory at Castle Street in Dromore left the building unstable when internal floors collapsed.
House	Carterton, New Zealand	Wood	2	May 27, 2002	Fire caused the center of the building to collapse
Apartment	Houston, TX	Unknown	2	May 26, 2002	Fire was contained to one building, where flames caused the roof to collapse.
Apartment	Owings Mills, MD	Unknown	3	April 13, 2002	Fire caused a partial collapse of the third floor.
Shopping	Elveden Forest, Suffolk, UK	Unknown	2	April 4, 2002	Early partial structural collapse
Apartment	New Orleans, LA	Wood	2	March 15, 2002	Fire damaged the attic of the building and caused a partial collapse of the floor in the apartment where the fire began.
Apartment over Business	Carthage, NY	Unknown	2	March 1, 2002	Fire resulted in heavy damage to at least four buildings, two collapsed.
Apartment	Dallas, TX	Unknown	2	February 10, 2002	Wall collapsed.
Grocery	Kenosha, WI	Unknown	2	January 1, 2002	Fire in a two-story corner grocery store building resulted in structural collapse.
House	Ottawa, Canada	Unknown	2	December 26, 2001	Fire destroyed the garage and reached the second story, eventually causing the roof to collapse
Apartment	Chicago, IL	Brick	3	December 6, 2001	A two- to three-foot concealed space between the old roof and the new third floor contributed to the rapid spread of the fire and led to the partial collapse.
House	South Wales, UK	Wood	2	November 26, 2001	The roof collapsed.
House	Wales, UK	Wood	2	November 26, 2001	Roof collapsed of a semi-detached home.
Duplex	Pittsburgh, PA	Wood	2	September 12, 2001	Fire caused floors to collapse.
House	Spotswood, NJ	Wood	2	August 11, 2001	Entire kitchen floor collapsed into the basement.

Table 2.4 Recent Low-Rise Building Fires With Collapses (Cont.)

Building Name or Occupancy	Location	Construction	Stories	Date	Nature and Extent of Fire
House	Pittsburgh, PA	Wood	2	August 8, 2001	Fire, located mainly on the second floor of the building, caused the roof of the home to collapse.
House	Irvine, CA	Unknown	2	July 6, 2001	Fire captain was injured after the burned-out second floor collapsed.
Apartment Row House	Passaic, NJ	Brick	3	May 9, 2001	Third floor collapse in building
	Baltimore, MD	Unknown	3	January 3, 2001	A man fell two or three stories when one floor of the burning row house gave way.
House	Smithfield, PA	Wood	2	September 12, 2000	Fire caused floors to collapse.
Store	Hopewell, VA	Unknown	2	June 4, 2000	Fire fueled by recycled clothes consumed a two-story building, which collapsed.
Apartment	Strathmore, UK	Unknown	3	April 8, 2000	Fire spread quickly to the roof area, causing the ceilings to collapse into the six adjoining units.
House	Stigler, Tulsa, OK	Unknown	2	February 15, 2000	House collapsed into the full basement.
House	Baltimore, MD	Unknown	3	December 24, 1999	Three-story frame house partially collapsed during the fire.
Country Club	Lewisburg, VA	Unknown	2	December 19, 1999	Fire apparently began on an open porch and spread through the attic of the 12,750-square foot, two-story building, resulting in the roof collapsing.
Apartment	Niagara Falls, NY	Brick	2.5	October 5, 1999	Fire spread quickly through the 2 1/2-story brick building and causing the roof and attic to collapse into the second floor
Apartment	San Francisco, CA	Unknown	3	July 20, 1999	Fire burned undetected in this area for a significant amount of time, eventually causing a second-floor collapse.
Apartment	Cunningham, Denver, CO	Unknown	3	July 4, 1999	Fire gutted the building's attic and caused the roof to collapse.
House	Cederburg, WI	Wood	2	April 29, 1999	Fire spread across the roof, and into the basement, causing the first floor in the kitchen and dining room to collapse.
House	Chingford, London, UK	Wood	2	March 6, 1999	Fire caused the roof and floors of the house in Bellamy Road to collapse.

Table 2.4 Recent Low-Rise Building Fires With Collapses (Cont.)

Building Name or Occupancy	Location	Construction	Stories	Date	Nature and Extent of Fire
Bar/Apartment	North Oakland, CA	Wood	2	January 10, 1999	Firefighters became trapped when the second floor of a nightclub collapsed during an interior fire attack.
House	Malden, Boston, MA	Wood	2	January 1, 1999	Large of fire related rubble caused the collapse of floor and ceiling, unable to determine the exact point of origin
Store	Glasgow, Scotland	Unknown	2	December 18, 1998	Fire caused the roof to collapse
House	Bayside, WI	Wood	2	October 31, 1998	Fire destroyed the home, causing the second floor to collapse.
Apartment/Commercial	Brooklyn, NY	Brick	3	June 5, 1998	Rear of the second floor of building #2 collapsed.
House	Baltimore, MD	Wood	3	March 24, 1998	Interior collapse of second and third floors.
House	Golden Valley, MN	Wood	3	March 12, 1998	Fire on all three levels resulted in partial wall collapses.
Apartment	Medford, MA	Wood	3	December 28, 1997	Partial collapse of second floor.
House	Toronto, Ontario, Canada	Unknown	2	November 21, 1997	Fire led to seconds floor collapse.
Row House	Pittsburgh, PA	Unknown	3	April 6, 1997	Fire weakened structure collapsed.
Apartment	Earlington, KY	Unknown	3	April 1, 1997	Top two floors collapsed.
Apartment	St. Henri, Montreal, Canada	Unknown	3	March 6, 1997	Walls began to collapse while firefighters were bringing the fire under control.
House	Stockton, CA	Wood	2	February 6, 1997	Second story floor toppled.
House	Jamaica, Queens, NY	Wood	2	December 31, 1996	Burn through of basement floor beams led to collapse of floor slabs.
House	Scituate, MA	Wood	2	September 15, 1996	Second-floor bedroom fire caused floor to collapse into the living room below.
House	Elliottsville, PA	Unknown	2	February 2, 1996	Two-story wood-frame house collapsed into foundation.
Store	Bangkok, Thailand	Unknown	3	November 27, 1995	Complete structural collapse
House	Edwardsville, MO	Wood	2	October 9, 1995	Fire gutted the attic of the two-story home and caused the roof to collapse.

Table 2.4 Recent Low-Rise Building Fires With Collapses (Cont.)

Building Name or Occupancy	Location	Construction	Stories	Date	Nature and Extent of Fire
Commercial/ Apartment	Los Angeles, CA	Wood	2	July 30, 1995	Burning two-story building collapsed.
Apartment	Calgary, Canada	Unknown	3	May 28, 1995	Ceiling on third floor started to collapse.
Apartment	West Palm Beach, FL	Unknown	2	April 21, 1995	Partial collapse of the second floor.
Apartment	Lynn, Boston, MA	Unknown	3	February 10, 1995	Roof and floors of three-story apartment building collapsed.
Warehouse	Seattle, WA	Unknown	2	January 5, 1995	Top floor collapsed onto lower floor, where the fire initiated.
Chalet	Granges-sur-Salvan, Switzerland	Wood	2	October 16, 1994	Progressive inward collapse of structure
Vacant	Washington, DC	Unknown	2	September 18, 1994	Exterior walls collapsed causing ceiling collapse.
Supply	Allentown, PA	Unknown	3	May 11, 1994	Total building collapse.
House	Roswell, GA	Wood	2	December 14, 1992	Total house collapse
Vacant	Lawrence, MA	Unknown	3	July 6, 1992	Total building collapse
Apartment	Brackenridge, PA	Concrete/unprotected steel	2	December 20, 1991	Section of the first-story floor assembly fell into the basement.

3.0 TASK 2. FIRE RESISTANCE TESTING LABORATORY SURVEY

3.1 Objective

The objective of this task was to survey the private and public facilities (for-profit, not-for-profit; academic, local, state and federal government; military and civilian; domestic and international) capable of testing the structural integrity of building elements under fire conditions. This survey was to determine their specific capabilities in this area and provide an understanding of the global capabilities for fire resistance testing.

3.2 Survey Scope and Methodology

The survey was limited to laboratories or facilities that could perform fire resistance testing. Fire resistance testing was defined as evaluating the structural integrity and/or the flame or temperature transmission through structural building elements. Structural building elements are defined as floors, roofs, ceilings, beams, columns, walls and their connections. Other test capabilities on fire performances characteristics such as flame spread, heat release rate, ignition resistance, smoke generation, etc. were not included in this survey.

The list of the test laboratories was developed based on personal knowledge, references from several of the laboratories, listings from various standards organizations (e.g. American Society for Testing and Materials (ASTM), International Standards Organization (ISO)) or approval organizations (e.g., International Conference of Building Officials (ICBO), Asia Pacific Laboratory Accreditation Cooperative (APLAC)). Table 3.1 provides a summary of the various laboratories and other contacts that were surveyed.

Each of the laboratories or contacts was initially contacted via E-mail and provided a questionnaire to complete and return. Follow-up contacts were performed as required.

The questionnaire was designed to provide a short yet complete description of the fire resistance testing capabilities of the various organizations. In general, the questionnaire requested information such as:

- Information on the location of the laboratory or facility
- Contact information
- Capability to test vertical building elements
    - What type
    - Size of elements / apparatus
    - Exposure conditions
    - Loading capabilities
- Capability to test horizontal building elements
    - What type
    - Size of elements / apparatus
    - Exposure conditions
    - Loading capabilities

- Information on restrictions for testing, i.e., for whom, environmental, etc.
- General pricing information
- Additional specialized facilities

An example of the questionnaire that was provided to the laboratories is provided in Appendix A.

3.3 Survey Responses

The various responses were collected and tabulated. In some cases, the contact or laboratory did not have existing fire resistance test facilities and thus they were eliminated from further consideration. In some cases, the laboratory or facility responded that that they were private facilities and did not accept outside testing and thus did not complete the form. Table 3.2 provides a summary of the responses received to date.

The responses from the laboratories that completed the questionnaire are summarized in Tables 3.3, 3.4, 3.5, and 3.6. Table 3.3 provides basic contact information for each laboratory. Tables 3.4 and 3.5 provide summaries of the test capabilities for vertical and horizontal elements, respectively. Table 3.6 provides additional information concerning testing restrictions and availability of other non-standard test capabilities. A copy of each of the completed questionnaires is provided in Volume II of this report.

3.4 Discussion of Survey

Currently, fire resistance testing is primarily performed to evaluate materials, products or assemblies with respect to their ability to maintain structural integrity or retard the passage of flames, or heat when exposed to a specified fire condition. The time that the test article exhibits this performance is known as its fire endurance or fire resistance rating. For example, a wall may attain a fire resistance rating of 1-hour, 2-hour or greater. This fire resistance rating, expressed in a time increment is typically specified in the building codes such that each structural building element for a particular building is required to have a rating that can vary from 0-hours to 4-hours or greater depending on the building requirements.

In order to attain a fire resistance rating, the material, product or assembly is evaluated using existing standard fire resistance tests. The most common fire resistance standards throughout the world are:

- ASTM E119 – "Standard Test Methods for Fire Tests of Building Construction and Materials"
- NFPA 251 – "Standard Methods of Tests of Fire Endurance of Building Construction and Materials"
- UL 263 – "Fire Tests of Building Construction and Materials"
- ISO 834 – "Fire Resistance Tests – Elements of Building Construction"

All of these tests are standardized fire resistance tests that have been used for many years. In general, they use similar fire exposure conditions as measured and controlled by the air temperature within the furnace. For some specific applications, a "hydrocarbon pool fire" exposure is also used. This test uses a heat flux to the test article as the exposure criteria but is controlled using the air temperature in the furnace. Typical "hydrocarbon pool fire" test methods followed in North America are:

- UL 1709 – "Rapid Rise Fire Tests of Protection Materials for Structural Steel"
- ASTM E 1529 – "Determining Effects of Large Hydrocarbon Pool Fires on Structural Members and Assemblies"

Figure 3.1 provides a plot of the fire exposure conditions for ASTM E119 (North America), ISO 834 (International) and the "hydrocarbon pool fire". All of the testing laboratories reported that they could perform the standard fire exposures (ASTM and ISO). Several of the laboratories can also perform fire exposures that are similar to the "hydrocarbon pool fire" exposure or greater.

Over the years, the fire test methods have standardized many of test details such as sample size, measurement of temperatures, loading, etc. As such, the majority of test laboratories have designed and built their current fire resistance test furnaces to meet the standard test methods.

The responding laboratories have various sized vertical and horizontal furnaces that can be used for fire resistance testing. In some cases, they also have small-scale furnaces that are used for scoping/R&D testing even though they do not meet the requirements of the standard tests with respect to the minimum size of the samples. In most of the laboratories, full-scale furnaces that can evaluate the standard sample sizes are available.

The standard practice in fire resistance testing is to test each building element individually. For example, roofs or floors are not tested in combination with walls or columns. Thus, the roof assembly is tested as an individual element and the column or the wall is also tested as an individual item. Connections between building elements are not necessarily evaluated under the existing fire resistance test methods. In some cases, such as with floors, the ends of beams may be restrained, but they are typically not exposed to the actual fire environment.

Full-scale vertical furnaces can typically evaluate samples approximately 3 m high x 3 m wide. The largest (ULC) can evaluate samples 4.6 m high x 4 m wide. Most vertical furnace are shallow (e.g. 0.6 m deep or less) but several (OPL, LGU) have depths up to 1.2 m to 1.5 m. Some of the test facilities can also provide expansion collars that will increase the depth of the furnace as well as the size (height & width) of the test sample.

Full-scale horizontal furnaces can evaluate samples up to approximately 4 m long x 5.5 m wide. The largest furnace (ULC) can accommodate a sample that is 10.5 m. long x 4 m. wide. Generally, the depth of the full-scale furnaces is approximately 1.8 m. The largest furnace described above, however, has a depth of 2.4 m. Most laboratories can build-up the sides of the horizontal furnaces such that deeper or taller samples, such as columns, can be tested.

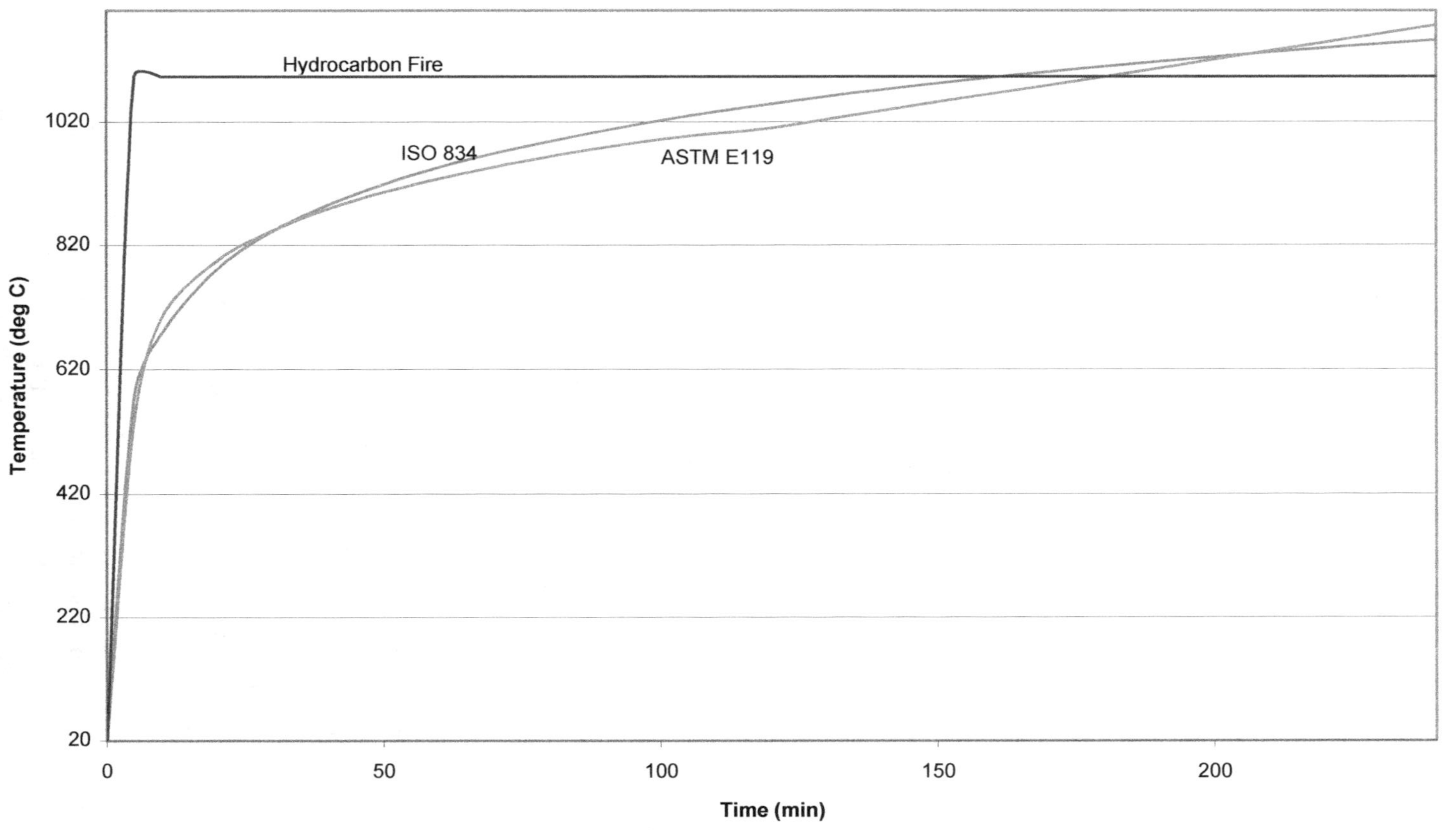

Figure 3.1 Fire Exposure Conditions

Testing combinations of structural elements can be conducted, but one must be careful with respect to the type and size of combinations. For example, it may be possible to test a full-scale floor and a column together within some limits on the size, but most laboratories cannot test a full-scale roof and wall combination. The horizontal furnace at ULC may have this capability, since they can shut down parts of the furnace and thus build a wall inside the furnace such that it will meet a floor. Also, OPL has a medium scale furnace that can potentially expose a wall (3 m tall x 3.6 m wide) in conjunction with a floor (3.6 m wide x 4.9 m long).

Loading of test assemblies is accomplished in various ways depending on the orientation of the test sample. Most of the laboratories load walls using hydraulic jacks such that the wall is in compression during the test. Columns are typically tested without load, since most of the testing involves fire protection materials on the columns and thus limiting temperatures are used rather than structural integrity. Several laboratories (NRC, LGU, BRI) have the capability to test columns under load.

In horizontal testing, loading is typically limited to live loads placed on top of the test assembly. The live loading is typically accomplished using water tanks or hydraulic jacks. Loading with respect to tension, shear, etc. is not typically done. One test laboratory (FPL) can perform tension loading in their small-scale horizontal furnace.

Generally, instrumentation in fire resistance testing is limited to:

- Temperatures / heat flux of exposing fire
- Temperatures on the test sample (interior and exterior)
- Deflection of the test sample
- Limited data on load history by some laboratories
- Pressure in the furnace with respect to atmosphere near the sample location

In some cases, research testing was conducted wherein load cells or strain gauges were used, but this is very rare. Some of the laboratories have the capability to use these devices but many experimental problems exist with their use. For example, at high temperatures these devices will be destroyed.

Most of the responding laboratories will perform testing for clients and there are few reported issues with respect to contracts and regulations. The largest concerns center on testing of potentially hazardous materials, such as asbestos.

The cost to perform fire resistance testing varies considerably depending on the type of assembly and its construction, instrumentation, loading, etc. In general, standard wall tests are in the range of $5,000 to $8,000 (US) while floor tests can range from $25,000 to $50,000 (US). Test scheduling is totally dependent upon the construction and instrumentation requirements for the test. For example, a simple wall test without the need for curing materials can be conducted within two weeks of receipt of materials while tests that may involve materials such as concrete may require a minimum of four to six months due to curing of the concrete.

Several laboratories have large, enclosed areas whereby specialized testing can be conducted. While not totally defined, several laboratories, (UL, SWR, SP) have facilities wherein a test structure can be constructed and potentially evaluated under real-scale fire conditions. While the overall space exists, the exact nature of the capabilities would be undefined until an actual proposed test is provided.

In summary, most of the responding laboratories have the capability to perform the standard fire resistance tests with respect to size, type of exposure, loading, and measurements. In some cases, laboratories can subject samples to a variety of fire exposures and several fairly large furnaces exist that may be used to evaluate, within some size limitations, connections or a combination of building elements. No single laboratory stands out as having all combinations of facilities such as largest size, capabilities of testing connections or combination of elements, loading, exposure conditions, etc. In general, the large commercial laboratories (e.g., ULC, UL, OPL, SWR, BRE, BRI, VTT) have greater than normal capabilities. The selection of a particular laboratory would depend on the specific objective of the test program and the test requirements. Real-scale size facilities and loading conditions and instrumentation other than those used or specified in the standard test methods is not be readily available.

Table 3.1 List of Laboratories/Contacts

Laboratory	Country
Underwriters Laboratories, Inc	USA
Underwriters Laboratories of Canada	Canada
Southwest Research Institute	USA
Intertek Testing – US & Canada	USA/Canada
Armstrong World Industries	USA
Factory Mutual Research Corporation	USA
SGS/US Testing Laboratory – LA	USA
Omega Point Laboratories	USA
3M Company	USA
Carboline	USA
National Research Council	Canada
US Forest Product Laboratory	USA
Western Fire Center	USA
US Gypsum – Research Facility	USA
Guardian	USA
NGL Testing Services	USA
VTEC Labs	USA
Commercial Testing	USA
UC Berkley	USA
Intertek – WI	USA
DITUC	Chile
IDIEM	Chile
INTI	Argentina
IPT	Brazil
REDCO	Belgium
Univ. of Ghent	Belgium
SP	Sweden
Danish Institute of Fire Technology	Denmark
VTT	Finland
SINTEF	Norway
Institute for QC of Bldg	Hungary
Lorient	UK
BRE	UK
Warrington	UK
Warrington	Australia
Lorient	Australia
CSIRO	Australia
Tianjin Fire Research Institute	China
China Nat. Center for QC & Testing of Bldg	China
Sichuan Fire Research Institute	China
Architecture & Building Research Institute	Taiwan
PSB/SISIR	Singapore
SIRIM	Malaysia
CTICM	France
CSTB	France
Building Res. Institute	Japan
Res. Institute of Marine Engineering	Japan
Gen Building Res. Corp	Japan
Japan Testing Center for Building Const.	Japan
Fire Research Institute	Czech Republic
Univ. of Canterbury	New Zealand
BRANZ	New Zealand
BAM	Germany
MPA – Materialprufungsamt	Germany
TNO	Holland

Table 3.2 Summary of Responses

Laboratory	Country	Capabilities	Remarks
Underwriters Laboratories, Inc	USA	Yes	
Underwriters Laboratories of Canada	Canada	Yes	
Southwest Research Institute	USA	Yes	
Armstrong World Industries	USA	Yes	Private
Factory Mutual Research Corporation	USA	None	
SGS/US Testing Laboratory – LA	USA	None	
Omega Point Laboratories	USA	Yes	
3M Compay	USA	Yes	Private
Carboline	USA	Yes	
National Research Council	Canada	Yes	
US Forest Produst Laboratory	USA	Yes	
US Gypsum – Research Facility	USA	Yes	
Guardian	USA	Yes	
NGL Testing Services	USA	Yes	
Commercial Testing	USA	None	
UC Berkley	USA	None	
DITUC	Chile	Yes	
IPT	Brazil	Yes	
Univ. of Ghent	Belgium	Yes	
VTT	Finland	Yes	
BRE	UK	Yes	
Warrington	UK	Yes	
CSIRO	Australia	Yes	
PSB/SISIR	Singapore	Yes	
SIRIM	Malaysia	Yes	
Building Res. Institute	Japan	Yes	
Res. Institute of Marine Engineering	Japan	Yes	
Gen Building Res. Corp	Japan	Yes	
SP Swedish National Testing and Res. Institute	Sweden	Yes	

Private = while some facilities exist, they did not wish to complete the form since they only do in-house testing.

Table 3.3 General Testing Laboratory Information

Laboratory	Location	Contact Person	Contact Information Tel: Fax: Email:	Code
PSB Corporation Pte Ltd	Singapore	Joseph Chng	(65) 6865 3778 (65) 6862 1433 joseph.chng@psbcorp.com	PSB
USG Research & Technology Center	Libertyville, IL, USA	Rich Kaczkowski	847-970-5255 847-970-5299	USG
Guardian Fire Testing Laboratory, Inc	Buffalo, NY, USA	R. Joseph Pearson	716-835-6880 716-835-5682 gftli@earthlink.net	GFT
Carboline Co.	Saint Louis, Mo, USA	Chris Magdalin	314-644-1000 314-644-4617 chris_Magdalin@Carboline.com	CBL
Laboratorio de Seguranca ao Fogo/ Instituto de Pesquisas Tecnologicas do Estado de Sao Paulo SA	Sao Paulo, Brazil	Antonio Fernando Berto	55 11 37674675 55 11 37674682 afberto@ipt.br	LSF
Laboratorio de Ensayo de Resistencia al Fuego DICTUC	Santiago, Chile	Pablo Matamala	(56 2) 686 4626 (56 2) 686 6226 pmatamal@ing.puc.cl	LER
Omega Point Laboratories, Inc	San Antonio, TX, USA	Deg Priest	210-635-8100 210-635-8101 dnpriest@ix.netcom.com	OPL
NGC Testing Services	Buffalo, NY, USA	Robert Menchetti	716-873-9750 ext.341 716-973-9753 email@ngctestingservices.com	NGC
NRC, Canada	Ottawa, Ontario, Canada	Mohamed A. Sultan	613-993-9771 613-954-0483 Mohamed.sultan@nrc.ca	NRC
Commercial Testing Co., Inc	Dalton, GA, USA	Jonathon Jackson	706-278-3935 706-278-3936 jjackson@commercialtesting.com	CTC
Research Institute of Marine Engineering	Tokyo, Japan	Tattsuhiro Hiraoka	81 42 394 3611 81 42 394 1119 hiraoka@rime.jp	RIM

Table 3.3 General Testing Laboratory Information (Cont.)

Laboratory	Location	Contact Person	Contact Information Tel: Fax: Email:	Code
USDA, Forest Service, Forest Products Laboratory	Madison, WI, USA	Robert White	608-231-9265 608-231-9508 rhwhite@fs.fed.us	FPL
Laboratory for Heat Transfer and Fuel Technology – Ghent Univ.	Ghent, Belgium	Prof dr ir P Vandevelde	32-9-243 77 55 32-9-243 77 51 paul.vandevelde@rug.ac.be	LGU
General Building Research Corporation of Japan	Osaka Prefecture, Japan	Masatomo Yoshido*	81 6872-0391 81 6872-0784 tasaka@gbrc.or.jp	GBR
BRE: Centre for Fire Performance and Suppression	Hertfordshire, UK	Dick Jones*	01923 665021 01923 665197 jonesr@bre.co.uk	BRE
VTT Technical Research Centre of Finland	Espoo, Finland	Matti Lanu	358 9 456 6935 358 9 456 4815 -	VTT
Underwriters Laboratory	Northbrook, IL, USA	Robert Berhinig	847-664-2292 847-509-6392 Robert.m.berhinig@us.ul.com	UL
Underwriters Laboratory Canada	Ontario, Canada	G. Abbas Nanji	416-757-3611 416-757-1781	ULC
Fire Engineering Testing Section, Testing Services Dept., SIRIM QAS Sdn. Bhd.	Selangor, Malaysia	Rohaya Ibrahim	603-5544 6465 603-5544 6454 rohaya@sirim.my	FET
Building Research Institute	Tsukuba, Japan	Dr. Mamoru Kohno	-	BRI
Fire Technology Laboratory, CSIRO Manufacturing and Infrastructure	Sydney, Australia	Garry E. Collins	61 2 9490 5408 61 2 94905528 Garry.Collins@csiro.au	FTL
Warrington Fire Research Centre	Cheshire, UK	Niall Rowan	44 1925 655116 44 1925 655419 niall.rowan@wfrc.co.uk	WFR
Southwest Research Institute, Dept. of Fire Technology	San Antonio, TX, USA	James R. Griffith	210-522-3716 210-522-3377 jgriffith@swri.org	SWR
SP Swedish National Testing and Res. Institute	Boras, Sweden	Ulf Wickstrom	+46 33 165194 ulf.wickstrom@sp.se	SP

* indicates one of multiple contact persons provided

Table 3.4 Vertical Structural Building Elements

Lab	Vertical FR Tests If Yes, Sizes	Combination Elements	Restrictions	Loading If Yes, How	Fire Exposure	Pressure (Pos., Neg., Combo)
USG	Max 10'x10'	No	No	Mechanical Jacks	Max Temp of 1200 °C	Yes
PSB	No					
GFT	3m x 3.7m	Yes but unwilling	No plastics or woods with arsenic	Hydraulic Jacks	ASTM E119	Yes
CBL	Columns – 1.5m only	Second Furnace Top Load	No	No	All	Column Furn. Pos. Toploand Furn Neg
LSF	2.8m x 2.8m	No	No	Axial Compression	ISO 834, ASTM E119	Yes
LER	Wall: 3.3m x 3.2m Columns: 3 m	Column-Beam Connection	Toxic Gases	Walls Only Jacks	ISO 834, ASTM E119	Yes
OPL	3m x 3m or 4.3m x 3.7m 2.7m columns	Yes	No	Hydraulics	All	Yes
NGC	3m x 3m	Yes	No	Hydraulic Jacks	ASTM E119	Yes
NRC	3m x 3.7m 1.2m x 1.8m	No	Toxic Materials	Hydraulic Jacks	All	Yes
CTC	No					
RIM	Walls 2.5m x 2.5m	No	High Comb. Mat.	Load Cells	ISO 834	Yes
FPL	2.4m x 3m	No	No	Hydraulic Jacks	ASTM E119	Yes but no ability to control pressure levels
LGU	3m x 3m	Yes	No	Load Cells	ASTM E119, ISO 834, Hydrocarbon	No
GBR	Yes	No	No	Load Beam and Jacks	ASTM E119, ISO 834, JIS A1304	Positive
BRE	3m x 3m	"potentially"	No	Hydraulic Jacks	ISO 834 ASTM E119 Any exposure	Yes Combo
VTT	3.2m x 3.2m	Wall-floor	No	Hydraulic Jacks	Any	No
UL	4.6m 3m	Yes	No	Hydraulic Jacks	ISO 834 ASTM E119	Yes per ISO834
ULC	4.6m x 4m (non-load bearing) 4.6m x 3.5m (load-bearing)	No	No	Hydraulic Jacks	ASTM E119 ISO 834	Combo
FET	Yes	No	No	Hydraulic Jack	BS476, Part 20	Combo with limitations
BRI	-					
FTL	3.5m x 3.5m	No	No	Compression only	Temp to 1400 deg C	Combo
WFR	1m x 1m and 3m x 3m	Possibly	No	Yes	ISO 834, ASTM E119, BS	Positive & Negative
SWR	3m x 3m	No	No	Yes	ISO 834, ASTM E119, and Pool Fire	Combo
SP	1.5 m x 1.5m and 4.3m x 4.3m	Yes	No	Live and Dead	Any	Positive & Negative
	3 m x 3 m	Yes	No	Compression	All	All

Table 3.5 Horizontal Structural Elements

Lab	Horizontal FR Tests If Yes, Sizes	Combination Elements	Restrictions	Loading If Yes, How	Fire Exposure	Pressure (Pos., Neg., Combo)
USG	2.3m x 2.6m	No	No	Water Containers	Max Temp of 1093°C	Yes
PSB	No					
GFT	4.2m x 5.4m	Yes but unwilling	No plastics, No arsenic	Water Tanks	ASTM E119 Max.4 hours	Yes
CBL	Columns – 1.5m only	No	No	No	All	No
LSF	No					
LER	4.05m x 4.8m	Column-Beam Connection	Toxic Gases	Dead Loads	ISO 834, ASTM E119	Yes
OPL	Min. Length of 3.7m	Yes	No	Hydraulics	All	Yes
NGC	4.3m x 5.5m	Yes	No	Water Containers	ASTM E119	Yes
NRC	4.1m x 4.6m and 1.2m x 1.8m	No	Toxic Materials	Hydraulic Jacks	All	Yes
CTC	No					
RIM	3m x 5m	No	High comb. Mat.	No	ISO 834	Yes
FPL	2.1m x 1.3m	No	No	Tensile Load Only	ASTM E119	No
LGU	6m x 3m x 2m	Yes	No	Hydraulic Jacks	ASTM E119, ISO834, and Pool Fire	Yes
GBR	2m x 7m x 1.5m	No	No	Loading Jacks	ASTM E119, ISO834, and Pool Fire	Positive
BRE	4m x 4m x 2m	Yes	No	Hydraulics	ASTM E119, ISO834, and Pool Fire	Positive
VTT	5m x 3m	Floor/Wall, T-beams and "cabins"	No	Hydraulic Jacks	Any	No
UL	4.3m x 5.3m	Yes	No	Water Tanks/Hydraulic Jacks	ASTM E119, ISO 834	Yes
ULC	10.5m x 4m x 2.4m	Yes	No	Jacks with dead weights	ASTM E119, ISO 834	Neg. or Pos.
FET	No					
BRI	4m x 8m	No	No	Compression only	Max Temp to 1400 °C	Combo
FTL	1m x 1m and 3.6m x 4.6m	Possibly	No	Dead Load on small Hydraulic Jacks	ISO 834, ASTM E119, and BS	Negative
WFR	4m x 3m	No	No	Dead Load & Hydraulic	ISO 834, ASTM E119, Pool Fire, and EN 1363-2	Combo
SWR	4.5m x 6m and 1.5 m x 1.5 m	Yes	No	Weights & Hydraulic Jacks	Any	Positive Negative
SP	5.2 m x 3.2 m	Yes	No	Hydraulic Jacks	All	All

Table 3.6 Additional Information

Lab	Outside Party Testing	Contractual Concerns	Turn-Around Time	Costs (US $)	Safety and Env. Regulations	Full-Scale Non-Std. Tests
USG	No		TBD	TBD		No
PSB	No	No	TBD	TBD	None	No
GFT	Yes	No	1 week	10,000 to 12,000	No plastics	No
CBL	No					No
LSF	Yes	No	2 months	2,000	None	No
LER	Yes	No	1 month	1,500 to 2,500	None	No
OPL	Yes	No	2 weeks	3,500 to 21,000	None	Multiple
NGC	Yes	No	2 weeks	5,000 to 10,000	None	No
NRC	Yes	No	Flexible	4,500 to 12,500	None	Yes
CTC	N/A	N/A	N/A	N/A	N/A	N/A
RIM	Yes	No	1 week	62,000	Air, water pollution	Yes
FPL	Cooperative projects	Gov't contracts	Limited staff	TBD	None	No
LGU	Yes	No	3-6 months	TBD	NR	Limited
GBR	Yes	No	NR	Varies	None	No
BRE	Yes	No	6-8 weeks	4,600 to 8,000	Risk assessment, method statement, liability insurance	Yes
VTT	Yes	No	3 months	6,000	None	Yes
UL	Yes	No	2 weeks	9,000	None	Yes
ULC	No	-	4weeks – 4months	9,000 to 30,000	None	Yes
FET	Yes	No	2-3 months	2,000 and up	No	No
BRI	Yes	"special contract"	3 months	4,000 to 8,000	No	Yes
FTL	Yes	Full disclosure	2-12 weeks	1,100 to 8,800	No	Yes
WFR	Yes	No	4-6 weeks	TBD	No	Some
SWR	Yes	No	4-6 weeks	5,000 to 10,000	No	Yes
SP	Yes	No	1-2 months	3,500 to 7,000	No	Yes

4.0 TASK 3. ASSESSMENT OF NEEDS

4.1 Objective

The objective of this task was to perform an assessment of the need for additional testing and/or experimental facilities to allow the performance of structural assemblies and fire resistance materials to be predicted under realistic fire conditions within actual buildings; and to develop options for meeting identified needs.

4.2 Assessment

The survey of the fire history of partial or total collapse of buildings identified 22 fire-induced collapses. These collapses when combined with fire events that caused major structural damage without collapse, result in a significant number of fire events. The costs, both economic and human, of these events are considerable. It is also apparent that significant damage or collapse occurs in all types of construction and is not limited to a particular building method or material. One must therefore, consider the entire range of building materials and construction methods as exhibiting the potential for fire-induced collapse.

One important finding is the lack of systematically collected information about partial or total collapse due to fire. Generally, very little structural collapse information was available, especially for events that occurred outside North America. Even in North America, there was no specific reporting of collapse via NFIRS, or other reporting system. Therefore, it is recommended that a structural damage and collapse reporting method be developed or be incorporated into existing databases whereby structural damage and collapse can be reported in the future. Without this information, the full scope of this problem cannot be understood.

Except for several recent research studies (Bailey, 2002; Newman, et al., 2000; Bailey, 2001) current fire resistance testing is limited to qualifying materials or assemblies using the standard fire resistance tests. When viewed from a whole building perspective, the standard fire testing has the following limitations:

- Tests are conducted on individual building elements of limited size.
- Tests do not evaluate the connections of one building element to another.
- Tests demonstrate the fire resistance capability based typically on limited temperature measurements and limited load carrying capability
- Other than significant deflections, no other structural measurements or data are obtained
- Tests typically use a similar but single fire exposure (i.e., ASTM E 199 and ISO 834)
- The test assembly is exposed from only one side, except for columns (i.e., all sides) and beams (i.e. typically 3 sides)
- Tests are typically not conducted to ultimate failure. Tests are stopped when the required rating period is attained, or when the appropriate temperature or deflection limit is observed.

This type of standardized fire resistance testing has been used for many years. It appears that it has provided some measure of safety, since most buildings that have a significant fire do not exhibit collapse. This may be due to many factors such as a limited fire spread area so that load is transferred to elements not exposed to the fire, redundancy and/or operation of fire protection features and over-design of the structural components, their fire protections and the other fire protection features within the building. However, despite these general successes, notable failures identified in this report have occurred. Based on the history, these failures will continue to occur.

Methods and tools need to be developed to provide an engineering basis for structural fire resistance. It is readily apparent that the current state-of-the-art in fire resistance testing does not address the "whole building" as a single entity. In this regard, the WTC collapses have highlighted the lack of communication between the structural engineering community and the fire protection community when designing buildings. Also, the issue of a single mode of failure (i.e., loss of some structural stability) resulting in collapse has arisen.

If a column is exposed to a fire, how its behavior or failure will affect the structural elements that it supports is unknown. Currently, we would test the column with protection such that the element will not exceed pre-defined critical temperatures below which its structural capability is assumed to be adequate. Even though the column is connected to a beam or floor, we do not evaluate the connection method, nor its protection, to assure that the connection will perform as designed. Also, if the column was to fail, the impact of this failure on the rest of the structure is unknown. It has always been assumed that if each individual element performs well by itself, then the whole will perform successfully. In some cases, this may be a faulty assumption.

It has been assumed in standard tests, that the actual fire will expose one side of an assembly. For example, a floor is tested with the fire under the floor. The connection of the floor to exterior columns or spandrel beams may be exposed both from the fire below and an exterior fire plume from window openings on the fire floor below. Thus, all of the elements are exposed and stressed simultaneously.

Structural performance at elevated temperatures except for the support of some live load or the attainment of a critical temperature, are typically ignored in current fire resistance testing. In effect, an assembly or element is evaluated for its structural performance by attainment of a critical temperature or its support of a live load based on some requirement for the amount of the load to be imposed during the test. In many cases, the live load used in the test may not be the maximum design load. Other loads or stresses may also be imposed on a structural element that may affect its structural performance during a fire event that may need to be considered. Thus, the combination of thermal load effects and structural load-carrying performance is ignored.

The existing fire test facilities evaluate building elements with specific size limitations such as length of beams (3.7 m), height of columns (2.7 m), etc. In real-world applications, building elements are significantly larger that those tested. This raises potential concerns of the structural performance on the large members when they are exposed to a fire. In these cases, the performance under a specific load on a shorter member will not reflect the same loading on a longer member.

In the standardized fire tests, the measurements and data collected are minimal. To understand how structural elements behave in a fire, greatly enhanced and unique instrumentation will be required. This will include monitoring the structural members for movement, state of stress, and temperature distribution within the elements. Without this information, it will be impossible to predict the performance of the building elements or systems when exposed to a fire.

The history of fire-induced building collapse coupled with identified concerns surrounding the existing fire resistance test methods clearly indicate that there is a need for reliable structural fire protection that goes beyond what is practiced today. In order to accomplish this goal of advancing our understanding and knowledge along with enhancing the combination of fire protection and structural engineering, empirical data developed from research-based, realistic tests must be generated. These data will be necessary to provide the basis for future analytical models and structural fire protection design methods.

Since our concerns center on a greater understanding of the impact of fire on the building elements, combination of building elements, their connections to each other and on real-scale size of building elements, most of our existing test facilities may be inadequate. Newer or unique facilities will be required to appropriately address these issues. If a connection, such as a column to a beam is to be evaluated, even the existing larger furnaces will have to be adapted to provide the appropriate exposure and loading systems. The current limitations on the size of the members that can be tested in existing furnaces remain a concern. In testing a combination of walls, exterior columns and floors, a unique test facility will have to be constructed to accommodate the required size, appropriate loading and the fire exposures needed. For longer, wider or taller members, a unique facility will again be required.

Currently, the path to the design, construction, and operation of a unique facility to provide the required research-based information is unclear. Future discussions within the fire protection and the structural engineering communities are required so as to begin to build a consensus with respect to the path forward. Deliberations with respect to developing a research plan for the future must continue and the plan developed. The Fire Resistance Determination and Performance Prediction Research Needs Workshop (Grosshandler, 2002) was a significant step in this direction. Along with the research plan, implementation strategies with respect to identification of stakeholders, investments of time and funding must be discussed and determined.

In summary, it is clear that to provide reliable structural fire protection for buildings, the future research needs will dictate that one or more unique facilities will be required. Specialized test apparatus, loading methods and instrumentation for large sizes, real-scale building elements and their connection, and full- or real-scale combinations of elements will be required as well as providing specialized and enhanced instrumentation to determine the structural performance of the building elements in a more quantitative manner. With the information and data that this type of unique facility can provide, both the fire protection and the structural engineering professions can be provided with an advanced understanding of the performance of fire exposed building elements such that improvements in the design and use of construction materials and techniques as well as the development of analytical models and design tools can be attained. This advanced knowledge can then be moved into actual design practice and future codes and standards so that fire-induced building collapse in buildings can be prevented.

5.0 REFERENCES

Abrams, M.S., "Behavior of Inorganic Materials in Fire," *ASTM Symposium on Design of Buildings for Fire Safety*, Boston, MA, ASTM Special Publication 685, June 1978.

Al-Mutairi, Naji M. and Al-Shaleh, Moneera S., "Assessment of Fire-Damaged Kuwaiti Structures," *Journal of Materials in Civil Engineering*, ASCE, Reston, VA, Feb. 1997, pp 7-14.

Bailey, C.G., "Holistic Behavior of Concrete Buildings in Fire," *Structures and Buildings*, **Vol. 152**, Issue 3, 2002.

Bailey, C.G., "Holistic Behavior of Concrete Buildings in Fire", pgs 199-212, *Proceedings of the Institution of Civil Engineers*, Structure and Buildings 152, Issue 3, August, 2002.

Bailey, C.G., "Steel Structures Supporting Composite Floor Slabs: Design for Fire", Building Research Establishment, 2001, BRE Digest 462.

Bell, Kim, "Squirrel Blamed in County Fire", *St. Louis Post-Dispatch*, Aug. 25, 1994, pg. 2B

Belser, Ann, "Two Men Trapped in Fire Refused to Try Jumping", *Pittsburgh Post-Gazette*, May 7, 2000, pg. B-2

Berto, Antonio Fernando and Tomina, Jose Carlos, "Lessons From the Fire in the CESP Administration Headquarters" IPT Building Technologies, Sao Paulo, Brazil, 1988.

Burke, John C. " One Person Dead, at Least Three Missing in Newton Blaze", *Associated Press*, Feb. 9, 2000.

Chui, Glenda, "Ingredients for An Inferno: Wind, Exposed Wood, No Sheetrock, Sprinkler System Not Yet Working", *The Mercury News*, Aug. 19, 2002.

Clark County Fire Department, "The MGM Grand Hotel Fire Investigation Report", 1981, Clark County, Nevada (available on Web site).

Evans, David, D., Mowrer, Frederick, W., "Progress Report on Fires Following the Northridge Earthquake" *Thirteenth Meeting of the UJNR Panel on Fire Research and Safety*, **Vol 2**, March 13-20, 1996, pp 271-360.

"Egyptian Prime Minister Visits Site of Factory Collapse", *Reuters News*, July 22, 2000.

EQE Web Site, Summary Disaster Reports on Loma Prieta, Northridge, and Kobe Earthquakes, www.EQE.com/publications/disasters.html .

"Factory Fire Kills 15 in Egypt", *BBC News, World: Asia-Pacific*, July 20, 2000

"Families Evacuated As Fire Destroys Bar", *The Herald (Glasgow)*, Feb. 27, 2001, pg. 4

FEMA 403, "World Trade Center Building Performance Study: Data Collection, Preliminary Observations, and Recommendations," FEMA, Washington, DC, May 2002.

"Fire Damages Portsmouth Nursing Home", Staff Report, *Portsmouth Times*, April 6, 1998, pg. B2.

Gathright, A., McCabe, M., and Rubenstein, S., "Fast-moving Inferno Destroys Upscale San Jose Development", *The San Francisco Chronicle*, August 20, 2002, pg. A1.

Hall, John R, Jr., "High-rise Building Fires," NFPA Report, Quincy, MA, September 2001.

Isner, Michael, S., "High-rise Office Building Fire, Alexis Nihon Plaza, Montreal Canada, October 26, 1986," NFPA Fire Investigation Report, Quincy, MA, 1986.

Klem, Thomas J., "First Interstate Bank Building Fire, Los Angeles, CA, May 4, 1988," NFPA Fire Investigation Report, Quincy, MA, 1988.

Klem, Thomas, J., "One Meridian Plaza, Philadelphia, PA, Three Firefighter Casualties, February 23, 1991," NFPA Fire Investigation Report, Quincy, MA, 1991.

Newman, G.M., Robinson, J.F., and Bailey, C.G., "Fire Safe Design: A New Approach to Multi-Story Steel-Framed Buildings", Steel Construction Institute, Berkshire, UK, 2000.

NFPA Fire Journal, "Collapse of the Hotel Vendome, Nine Firefighters Killed," Quincy, MA, January 1973, pp 34-41.

"One New York Plaza Fire," Report by the New York Board of Fire Underwriters, Bureau of Fire Prevention and Public Relations, New York, NY, 1970.

Onishi, Norimitsu, "Fire in Bronx Apartments Kills 3, Including Woman and Son", *New York Times*, April 6, 1994, pg. B-3.

Papaioannou, Kyriakos, "The Conflagration of Two Large Department Stores in the Centre of Athens," *Fire and Materials*, **Vol. 10**, John Wiley and Sons, Ltd.,1986, pg 171-177.

"Russia: One Dead, 400 Homeless After Building Collapses", *The Ottawa Citizen*, CanWest Global Communications Corp., pg. A18, June 4, 2002.

"Russian Apartment Block Collapses", BBC News Online: World:Europe, June 3, 2002.

Sharry, John A., Culver, Charles, Crist, Robert, and Hillelson, Jeffrey, "Military Personnel Records Center Fire," *NFPA Fire Journal,* Quincy, MA, May 1974.

Stepan, Cheryl, "I Thought the Worst was Over; Firefighers Buried for 30 Minutes", *The Hamilton Spectator*, Feb, 9, 2002, pg. A-03.

Sugahara, Shin'Ichi, "Building Firesafety Design against a Large Earthquake-Based on the 1995 Kobe-Hanshin Earthquake

Tong, Betsy, Q.M., "Victims of Blaze Make do at Shelter", Contributing Reporter, Metro/Region Section (pg. 14) of Cambridge News, City Edition, Oct. 4, 1993.

Willey, Elwood A., "High-rise Building Fire, Sao Paulo, Brazil," *NFPA Fire Journal*, Quincy, MA, July 1972, pp 6-13, 105-109.

APPENDIX A

SAMPLE FIRE RESISTANCE QUESTIONNAIRE

Questionnaire Concerning Fire Resistance Testing
Of Structural Building Elements

PART I – GENERAL

Name of Laboratory

Location – City, State, Country

Type of Organization – Government, Private, Industry, Non-profit, etc.

Contact information – Person, address, E-mail address, phone number, and fax number.

Can the laboratory perform fire resistance testing of structural building elements? These include walls, columns, floors, ceilings, beams, trusses, connections etc.

If no – so state, complete Part I and return the form to Hughes Associates, Inc. as shown in Section V at the bottom of this Questionnaire.

If yes – please continue and address the parts that are applicable to your laboratory.

PART II – VERTICAL STRUCTURAL BUILDING ELEMENTS

Can you test for fire resistance of vertical building elements such as walls/columns, etc.?

Can you perform standard fire tests and nonstandard fire resistance tests on these elements?

How many furnaces for vertical element tests do you have and what elements can be tested in each?

Besides testing discrete building elements such as walls or columns by themselves, can your furnace(s) accommodate combinations of elements such as wall-floor combinations, column-beam connections, etc.?

What are the largest dimensions of test samples that you can test in each furnace for each type or combination of building element?

Do you have any restrictions with respect to what type of materials can be tested (i.e., wood, steel, concrete, composites, aluminum, fire-proofing materials, etc.)?

What type for fire exposure can your furnace(s) generate (i.e., ISO 834, ASTM E119, Hydrocarbon pool fire simulations, others)?

Do you have any limitations on the fire exposure?

Do you control the furnace exposure via temperatures and/or heat flux?

How are these parameters measured (i.e., thermocouples, heat flux gauges, plate thermometers, etc.) and recorded (i.e., computer, strip charts, etc.)?

Can your furnace(s) run positive, negative or a combination of both with respect to pressure inside the furnace? Do you have any limitations with respect to the pressure or the position of the neutral plane?

Do you measure and record the furnace pressure?

Can you apply a load to the test sample?

If so, can this be done on each furnace?

Provide a brief description of the load mechanism (i.e., compression, tension, shear, axial, load limits, use load beam w/ jacks, etc.) for each furnace and/or type of building element.

How is the load carrying capability monitored during the test?

PART III – HORIZONTAL STRUCTURAL BUILDING ELEMENTS

Can you test for fire resistance of horizontal building elements such as floors, ceilings, beams, trusses, etc.?

Can you perform standard fire tests and nonstandard fire resistance tests on these elements?

How many furnaces for horizontal element tests do you have and what elements can be tested in each?

Besides testing discrete building elements such as floors or beams by themselves, can your furnace(s) accommodate combinations of elements such as floor-wall combination, column-beam connections, etc.?

What are the largest dimensions of test samples that you can test in each furnace for each type or combination of building element?

Do you have any restrictions with respect to what type of materials can be tested (i.e., wood, steel, concrete, composites, aluminum, fire-proofing materials, etc.)?

What type for fire exposure can your furnace(s) generate (i.e., ISO 834, ASTM E119, Hydrocarbon pool fire simulations, others)?

Do you have any limitations on the fire exposure?

Do you control the furnace exposure via temperatures and/or heat flux?

How are these parameters measured (i.e., thermocouples, heat flux gauges, plate thermometers, etc.) and recorded (i.e., computer, strip charts, etc.)?

Can your furnace(s) run positive or negative with respect to pressure inside the furnace? Do you have any limitations with respect to the pressure or the position of the neutral plane?

Do you measure and record the furnace pressure?

Can you apply a load to the test sample?

If so, can this be done on each furnace?

Provide a brief description of the load mechanism (i.e., compression, tension, shear, axial, load limits, dead load, live load, use loading jacks/frame, water containers, etc.) for each furnace and/or type of building element.

How is the load carrying capability monitored during the test?

PART IV – ADDITIONAL INFORMATION

Is your laboratory capable of and willing to performing tests for outside parties or do you test only for yourself or your industry?

Do you have any restrictions on doing testing for outside parties?

Are there any specific contractual concerns with respect to performing testing for outside parties?

What is the typical turn-around time for conducting tests? We realize that cure times, construction, etc. may affect this but in general what is the timing.

Could you please provide a range of costs for some typical tests such as walls, columns, or floors? We realize many factors will influence this, but we are only looking for some general cost information.

Besides normal safety and environmental regulations, does your laboratory have any special regulations or rules concerning such items as testing, materials or disposal? If so, briefly details these.

Besides, the furnaces described above, does your laboratory have the capability to perform full-scale, nonstandard, fire resistance tests of structures such as rooms or small buildings?

Any other information that may assist us in our survey.

SECTION V – SUBMITTAL INFORMATION

Feel free to answer each question using this form and then submit the completed form.

This form can be submitted via Fax or E-mail. Please submit the completed form to:

Jesse J. Beitel
Senior Scientist
Hughes Associates, Inc.
3610 Commerce Drive, Suite 817
Baltimore, Maryland 21227 USA

Phone – 410-737-8677
Fax – 410-737-8688

E-mail – jbeitel@haifire.com

Thank you for your assistance and time, and if you have any questions, please feel free to contact me.

APPENDIX B

SAMPLES* OF LABORATORY QUESTIONNAIRES

Code	Laboratory	Page
LER	Laboratorio de Ensayo de Resistencia al Fuego DICTUC	56
LGU	Laboratory for Heat Transfer and Fuel Technology – Ghent Univ.	62
NGC	NGC Testing Services	66
PSB	PSB Corporation Pte Ltd	70
UL	Underwriters Laboratory	75
USG	USG Research & Technology Center	82

Permission obtained from listed laboratories to release details of survey. Contact other laboratories directly to request their additional information.

LER

PART I – GENERAL

Name of Laboratory:

Laboratorio de Ensayo de Resistencia al Fuego
DICTUC

Location – City, State, Country:

Santiago, Chile

Type of Organization – Government, Private, Industry, Non-profit, etc.

Private. It is owned by the Catholic University of Chile (non profit).

Contact information – Person, address, E-mail address, phone number, and fax number:

Pablo Matamala P.
Vicuña Mackenna 4860, Macul, Santiago, Chile
pmatamal@ing.puc.cl
Phone number: (56 2) 686 4626 – 686 4265
Fax number: (56 2) 686 6226

Can the laboratory perform fire resistance testing of structural building elements? These include walls, columns, floors, ceilings, beams, trusses, connections etc.

Yes.

If no – so state, complete Part I and return the form to Hughes Associates, Inc. as shown in Section V at the bottom of this Questionnaire.

If yes – please continue and address the parts that are applicable to your laboratory.

PART II – VERTICAL STRUCTURAL BUILDING ELEMENTS

Can you test for fire resistance of vertical building elements such as walls/columns, etc.?

Yes.

Can you perform standard fire tests and nonstandard fire resistance tests on these elements?

Yes.

How many furnaces for vertical element tests do you have and what elements can be tested in each?

We have one furnace, capable of performing test for walls and columns.

Besides testing discrete building elements such as walls or columns by themselves, can your furnace(s) accommodate combinations of elements such as wall-floor combinations, column-beam connections, etc.?

We have not performed such a test, but a column – beam connection could be accommodated.

What are the largest dimensions of test samples that you can test in each furnace for each type or combination of building element?

Walls: 3,3 meters high and 3,2 meters wide (3,1 meters x 3,0 meters exposed to the fire)
Columns: 3 meters long (2,7 meters exposed to the fire)

Do you have any restrictions with respect to what type of materials can be tested (i.e., wood, steel, concrete, composites, aluminum, fire-proofing materials, etc.)?

We do not have any restriction. The only concern would be the emission of dangerous substances: we have an incinerator to burn out the combustion gases, keeping them at 850°C during 2 seconds at least.

What type for fire exposure can your furnace(s) generate (i.e., ISO 834, ASTM E119, Hydrocarbon pool fire simulations, others)?

ISO 834 and ASTM E119. Hydrocarbon pool could be done, but we have not tested it yet.

Do you have any limitations on the fire exposure?

Hydrocarbon pool could be done, but we have not tested it yet. It would depend on the type of testing specimen: probably with a column we would not have problems, but regarding a wall we are not sure yet.

Do you control the furnace exposure via temperatures and/or heat flux?

We control it via temperature.

How are these parameters measured (i.e., thermocouples, heat flux gauges, plate thermometers, etc.) and recorded (i.e., computer, strip charts, etc.)?

Thermocouples and we are going to implement plate thermometers as well.
We record all the information via a computer. We read temperatures every one second.

Can your furnace(s) run positive, negative or a combination of both with respect to pressure inside the furnace? Do you have any limitations with respect to the pressure or the position of the neutral plane?

We can run positive, negative or a combination of both. We have not run it to the limits: we can get full possitive pressure, but we have not tested the maximum height of the neutral plane.

Do you measure and record the furnace pressure?

We measure it and then record it via a computer. We can read the pressure every 1 second.

Can you apply a load to the test sample?

Only to walls; not columns.

If so, can this be done on each furnace?

Yes. (We have only one furnace.)

Provide a brief description of the load mechanism (i.e., compression, tension, shear, axial, load limits, use load beam w/ jacks, etc.) for each furnace and/or type of building element.

For walls only: It is done in compression using an external frame and jacks. We can apply up to 30 tons of load.

How is the load carrying capability monitored during the test?

It is read from the jacks (related to the oil pressure).

PART III – HORIZONTAL STRUCTURAL BUILDING ELEMENTS

Can you test for fire resistance of horizontal building elements such as floors, ceilings, beams, trusses, etc.?

Yes.

Can you perform standard fire tests and nonstandard fire resistance tests on these elements?

Yes.

How many furnaces for horizontal element tests do you have and what elements can be tested in each?

We have only one furnace. We can test floors, ceilings and floor – beams arrangements.

Besides testing discrete building elements such as floors or beams by themselves, can your furnace(s) accommodate combinations of elements such as floor-wall combination, column-beam connections, etc.?

We can arrange and test column – beam arrangements.

What are the largest dimensions of test samples that you can test in each furnace for each type or combination of building element?

The fire exposed surface of the element is 4,5 meters by 3,75 meters. The element should be of 4,05 meters x 4,8 meters to fit into the frame.

Do you have any restrictions with respect to what type of materials can be tested (i.e., wood, steel, concrete, composites, aluminum, fire-proofing materials, etc.)?

We do not have any restriction. The only concern would be the emission of dangerous substances: we have an incinerator to burn out the combustion gases, keeping them at 850ºC during 2 seconds at least.

What type for fire exposure can your furnace(s) generate (i.e., ISO 834, ASTM E119, Hydrocarbon pool fire simulations, others)?

ISO 834 and ASTM E119. Hydrocarbon pool could be done, but we have not tested it yet.

Do you have any limitations on the fire exposure?

Hydrocarbon pool could be done, but we have not tested it yet. It would depend on the type of testing specimen: probably with a column we would not have problems, but regarding a wall we are not sure yet.

Do you control the furnace exposure via temperatures and/or heat flux?

We control it via temperature.

How are these parameters measured (i.e., thermocouples, heat flux gauges, plate thermometers, etc.) and recorded (i.e., computer, strip charts, etc.)?

Thermocouples and we are going to implement plate thermometers as well.
We record all the information via a computer. We read temperatures every one second.

Can your furnace(s) run positive or negative with respect to pressure inside the furnace? Do you have any limitations with respect to the pressure or the position of the neutral plane?

We can run positive, negative or a combination of both. We have not run it to the limits: we can get full positive pressure, but we have not tested the maximum height of the neutral plane.

Do you measure and record the furnace pressure?

We measure it and then record it via a computer. We can read the pressure every 1 second.

Can you apply a load to the test sample?

Yes.

If so, can this be done on each furnace?

Yes. (We have only one furnace.)

Provide a brief description of the load mechanism (i.e., compression, tension, shear, axial, load limits, dead load, live load, use loading jacks/frame, water containers, etc.) for each furnace and/or type of building element.

We put dead loads over the test specimen. The maximum capacity is 500 Kg/m^2.

How is the load carrying capability monitored during the test?

We use dead load; they are weighted before. Deflection is monitored during the test.

PART IV – ADDITIONAL INFORMATION

Is your laboratory capable of and willing to performing tests for outside parties or do you test only for yourself or your industry?

We perform tests for outside parties. We are also available to perform tests for the university for investigation purposes.

Do you have any restrictions on doing testing for outside parties?

No.

Are there any specific contractual concerns with respect to performing testing for outside parties?

No.

What is the typical turn-around time for conducting tests? We realize that cure times, construction, etc. may affect this but in general what is the timing.

After the specimen is finished, the test is performed in one day and the report is available two weeks after this.

For example: we tested some cardboard/gypsum and glass fiber arranged partition walls: they were built in one day each, and 3 days after letting them dry, we run the test. We could perform 3 tests in one week.

A horizontal floor assembly had to be built in 2 days, cured in 28 days and tested in two days.

Could you please provide a range of costs for some typical tests such as walls, columns, or floors? We realize many factors will influence this, but we are only looking for some general cost information.

Vertical Elements: US$ 1500 - 2000
Horizontal Elements: US$ 2000 - 2500

These prices do not consider the building costs of the specimen (materials and work hours)

Besides normal safety and environmental regulations, does your laboratory have any special regulations or rules concerning such items as testing, materials or disposal? If so, briefly details these.

No.

Besides, the furnaces described above, does your laboratory have the capability to perform full-scale, nonstandard, fire resistance tests of structures such as rooms or small buildings?

No.

Any other information that may assist us in our survey.

No.

LGU

PART I – GENERAL

Name of Laboratory

LABORATORY FOR HEAT TRANSFER AND FUEL TECHNOLOGY – GHENT UNIVERSITY

Location – City, State, Country

OTTERGEMSESTEENWEG 711, 9000 GHENT, BELGIUM

Type of Organization – Government, Private, Industry, Non-profit, etc.

GOVERNMENT

Contact information – Person, address, E-mail address, phone number, and fax number.

PROF DR IR P VANDEVELDE, LABORATORY FOR HEAT TRANSFER AND FUEL TECHNOLOGY – GHENT UNIVERSITY, OTTERGEMSESTEENWEG 711, 9000 GHENT, BELGIUM, PAUL.VANDEVELDE@RUG.AC.BE, TEL. +32-9-243 77 55, FAX +32-9-243 77 51

Can the laboratory perform fire resistance testing of structural building elements? These include walls, columns, floors, ceilings, beams, trusses, connections etc.
YES

If no – so state, complete Part I and return the form to Hughes Associates, Inc. as shown in Section V at the bottom of this Questionnaire.

If yes – please continue and address the parts that are applicable to your laboratory.

PART II – VERTICAL STRUCTURAL BUILDING ELEMENTS

Can you test for fire resistance of vertical building elements such as walls/columns, etc.? **YES**

Can you perform standard fire tests and nonstandard fire resistance tests on these elements? **YES**

How many furnaces for vertical element tests do you have and what elements can be tested in each? **1 – Doors, walls, loaded and unloaded columns, lift landing doors, glazed elements, linear gap seals, dampers.**

Besides testing discrete building elements such as walls or columns by themselves, can your furnace(s) accommodate combinations of elements such as wall-floor combinations, column-beam connections, etc.? **To a limited extent.**

What are the largest dimensions of test samples that you can test in each furnace for each type or combination of building element? **The furnace dimensions are 3 m width, 3 m height and ± 1,5 m depth for doors, walls, lift landing doors, glazed elements, linear gap seals and dampers. The furnace dimensions are 3 m width, ± 4 m height and 3 m depth for loaded and unloaded columns.**

Do you have any restrictions with respect to what type of materials can be tested (i.e., wood, steel, concrete, composites, aluminum, fire-proofing materials, etc.)? **No.**

What type for fire exposure can your furnace(s) generate (i.e., ISO 834, **YES** ASTM E119 **YES**, Hydrocarbon pool fire simulations **NO**, others **YES, hydrocarbon curve**)?

Do you have any limitations on the fire exposure? **Six hours duration.**

Do you control the furnace exposure via temperatures and/or heat flux?
Thermocouple plus plate thermometers (European standards).
How are these parameters measured (i.e., <u>thermocouples</u>, heat flux gauges, <u>plate thermometers</u>, etc.) and recorded (i.e., <u>computer</u>, strip charts, etc.)?

Can your furnace(s) run positive, negative or a combination <u>of both with respect to pressure inside the furnace</u>? Do you have any limitations with respect to the pressure or the position of the neutral plane? **The neutral plane within the furnace height if the pressure is negative in the furnace. The pressure in the furnace is up to circa 50 Pa.**

Do you measure and record the furnace pressure? **YES**

Can you apply a load to the test sample? **YES**

If so, can this be done on each furnace? **YES**

Provide a brief description of the load mechanism (i.e., compression, tension, shear, axial, load limits, use load beam w/ jacks, etc.) for each furnace and/or type of building element. **We can charge loaded columns up to 5.000 kN.**

How is the load carrying capability monitored during the test? **Pressure of hydraulic rams and/or load cells.**

PART III – HORIZONTAL STRUCTURAL BUILDING ELEMENTS

Can you test for fire resistance of horizontal building elements such as floors, ceilings, beams, trusses, etc.? **YES**

Can you perform standard fire tests and nonstandard fire resistance tests on these elements? **YES**

How many furnaces for horizontal element tests do you have and what elements can be tested in each? **One. Floors, beams, trusses, ceilings, ventilation ducts, stairs, columns.**

Besides testing discrete building elements such as floors or beams by themselves, can your furnace(s) accommodate combinations of elements such as floor-wall combination, column-beam connections, etc.? **To a limited extent.**

What are the largest dimensions of test samples that you can test in each furnace for each type or combination of building element? **Furnace = 6 m long, 3 m wide, 2 m deep.**

Do you have any restrictions with respect to what type of materials can be tested (i.e., wood, steel, concrete, composites, aluminum, fire-proofing materials, etc.)? **NO**

What type for fire exposure can your furnace(s) generate (i.e., ISO 834 **YES**, ASTM E119 **YES**, Hydrocarbon pool fire simulations, others **YES**)?

Do you have any limitations on the fire exposure? **Duration < 6 hours**

Do you control the furnace exposure via temperatures and/or heat flux?
As above.

How are these parameters measured (i.e., thermocouples, heat flux gauges, plate thermometers, etc.) and recorded (i.e., computer, strip charts, etc.)?
As above.
Can your furnace(s) run positive or negative with respect to pressure inside the furnace? Do you have any limitations with respect to the pressure or the position of the neutral plane?
As above.
Do you measure and record the furnace pressure? **YES**

Can you apply a load to the test sample? **YES**

If so, can this be done on each furnace? **YES**

Provide a brief description of the load mechanism (i.e., compression, tension, shear, axial, load limits, dead load, live load, use loading jacks/frame, water containers, etc.) for each furnace and/or type of building element. **We can charge beams up to 800 kN.**

How is the load carrying capability monitored during the test? **As above.**

PART IV – ADDITIONAL INFORMATION

Is your laboratory capable of and willing to performing tests for outside parties or do you test only for yourself or your industry? **YES**

Do you have any restrictions on doing testing for outside parties? **NO**

Are there any specific contractual concerns with respect to performing testing for outside parties? **Normal commercial conditions.**

What is the typical turn-around time for conducting tests? We realize that cure times, construction, etc. may affect this but in general what is the timing.

Three to six months, depending on complexity and availability of furnace.

Could you please provide a range of costs for some typical tests such as walls, columns, or floors? We realize many factors will influence this, but we are only looking for some general cost information.

Besides normal safety and environmental regulations, does your laboratory have any special regulations or rules concerning such items as testing, materials or disposal? If so, briefly details these.

Besides, the furnaces described above, does your laboratory have the capability to perform full-scale, nonstandard, fire resistance tests of structures such as rooms or small buildings? **To a limited extent. The tests are done in the open air.**

Any other information that may assist us in our survey.

NGC

PART I – GENERAL

NGC Testing Services

1650 Military Road, Buffalo, NY 14217

Private

Robert J. Menchetti, email@ngctestingservices.com, 716 8739750 ext. 341, 716 973-9753 (fax)

Can the laboratory perform fire resistance testing of structural building elements? These include walls, columns, floors, ceilings, beams, trusses, connections etc.   Yes

If no – so state, complete Part I and return the form to Hughes Associates, Inc. as shown in Section V at the bottom of this Questionnaire.

If yes – please continue and address the parts that are applicable to your laboratory.

PART II – VERTICAL STRUCTURAL BUILDING ELEMENTS

Can you test for fire resistance of vertical building elements such as walls/columns, etc.? Yes

Can you perform standard fire tests and nonstandard fire resistance tests on these elements? Yes

How many furnaces for vertical element tests do you have and what elements can be tested in each?   1 vertical furnace, 3 test movable 10'x10' frames

Besides testing discrete building elements such as walls or columns by themselves, can your furnace(s) accommodate combinations of elements such as wall-floor combinations, column-beam connections, etc.?   Yes

What are the largest dimensions of test samples that you can test in each furnace for each type or combination of building element?   10' x 10'

Do you have any restrictions with respect to what type of materials can be tested (i.e., wood, steel, concrete, composites, aluminum, fire-proofing materials, etc.)? NO

What type for fire exposure can your furnace(s) generate (i.e., ISO 834, ASTM E119, Hydrocarbon pool fire simulations, others)? E119

Do you have any limitations on the fire exposure? NO

Do you control the furnace exposure via temperatures and/or heat flux? Temp.

How are these parameters measured (i.e., thermocouples, heat flux gauges, plate thermometers, etc.) and recorded (i.e., computer, strip charts, etc.)? Computer

Can your furnace(s) run positive, negative or a combination of both with respect to pressure inside the furnace? Do you have any limitations with respect to the pressure or the position of the neutral plane? Yes, no limitations

Do you measure and record the furnace pressure? Yes

Can you apply a load to the test sample? Yes

If so, can this be done on each furnace? Yes

Provide a brief description of the load mechanism (i.e., compression, tension, shear, axial, load limits, use load beam w/ jacks, etc.) for each furnace and/or type of building element. Axial load utilizing hydraulic jacks

How is the load carrying capability monitored during the test? Pressure gauges / deflection gauges

PART III – HORIZONTAL STRUCTURAL BUILDING ELEMENTS

Can you test for fire resistance of horizontal building elements such as floors, ceilings, beams, trusses, etc.? YES

Can you perform standard fire tests and nonstandard fire resistance tests on these elements? YES

How many furnaces for horizontal element tests do you have and what elements can be tested in each? 2

Besides testing discrete building elements such as floors or beams by themselves, can your furnace(s) accommodate combinations of elements such as floor-wall combination, column-beam connections, etc.? YES

What are the largest dimensions of test samples that you can test in each furnace for each type or combination of building element? 14' x 18'

Do you have any restrictions with respect to what type of materials can be tested (i.e., wood, steel, concrete, composites, aluminum, fire-proofing materials, etc.)? NO

What type for fire exposure can your furnace(s) generate (i.e., ISO 834, ASTM E119, Hydrocarbon pool fire simulations, others)? E119

Do you have any limitations on the fire exposure? NO

Do you control the furnace exposure via temperatures and/or heat flux? Temp.

How are these parameters measured (i.e., thermocouples, heat flux gauges, plate thermometers, etc.) and recorded (i.e., computer, strip charts, etc.)? Computer

Can your furnace(s) run positive or negative with respect to pressure inside the furnace? Do you have any limitations with respect to the pressure or the position of the neutral plane? Yes, no limitations

Do you measure and record the furnace pressure? YES

Can you apply a load to the test sample? YES

If so, can this be done on each furnace? YES

Provide a brief description of the load mechanism (i.e., compression, tension, shear, axial, load limits, dead load, live load, use loading jacks/frame, water containers, etc.) for each furnace and/or type of building element. WATER CONTAINERS

How is the load carrying capability monitored during the test? Deflection gauges

PART IV – ADDITIONAL INFORMATION

Is your laboratory capable of and willing to performing tests for outside parties or do you test only for yourself or your industry? We perform tests for outside parties.

Do you have any restrictions on doing testing for outside parties? NO

Are there any specific contractual concerns with respect to performing testing for outside parties? NO

What is the typical turn-around time for conducting tests? We realize that cure times, construction, etc. may affect this but in general what is the timing. Within 2 weeks after receiving test materials

Could you please provide a range of costs for some typical tests such as walls, columns, or floors? We realize many factors will influence this, but we are only looking for some general cost information. Walls: $ 5,000 +/- , Floor – ceiling +/- $ 10,000.

Besides normal safety and environmental regulations, does your laboratory have any special regulations or rules concerning such items as testing, materials or disposal? If so, briefly details these. NO

Besides, the furnaces described above, does your laboratory have the capability to perform full-scale, nonstandard, fire resistance tests of structures such as rooms or small buildings?   NO, not beyond room corner burns

Any other information that may assist us in our survey.

Our web site: ngctestingservices.com

We also have a full building acoustics laboratory within the same facility

PSB

PART I – GENERAL

Name of Laboratory: PSB Corporation Pte Ltd

Location – No 10 Tuas Avenue 10 Singapore 639134

Type of Organization – Government Linked Company

Contact information – Joseph Chng
Tel: (65) 6865 3778
Fax:(65) 6862 1433
Email:

Can the laboratory perform fire resistance testing of structural building elements? These include walls, columns, floors, ceilings, beams, trusses, connections etc.

Yes, only on non-load bearing elements.

If no – so state, complete Part I and return the form to Hughes Associates, Inc. as shown in Section V at the bottom of this Questionnaire.

If yes – please continue and address the parts that are applicable to your laboratory.

PART II – VERTICAL STRUCTURAL BUILDING ELEMENTS

Can you test for fire resistance of vertical building elements such as walls/columns, etc.?

Yes

Can you perform standard fire tests and nonstandard fire resistance tests on these elements?

Yes

How many furnaces for vertical element tests do you have and what elements can be tested in each?

2 number of vertical furnaces of size 3.2m x 3.2m and 1 number of cube furnace of size 1.5m x 1.5m, for testing of wall partition, fire doors, dampers, lift landing doors, refute chute hopper,etc.

Besides testing discrete building elements such as walls or columns by themselves, can your furnace(s) accommodate combinations of elements such as wall-floor combinations, column-beam connections, etc.?

No

What are the largest dimensions of test samples that you can test in each furnace for each type or combination of building element?

3m x 3m

Do you have any restrictions with respect to what type of materials can be tested (i.e., wood, steel, concrete, composites, aluminum, fire-proofing materials, etc.)?

No

What type for fire exposure can your furnace(s) generate (i.e., ISO 834, ASTM E119, Hydrocarbon pool fire simulations, others)?

ISO 834 , ASTM E119, BS 476 Part 20/22, AS 1530.4

Do you have any limitations on the fire exposure?

No

Do you control the furnace exposure via temperatures and/or heat flux?

Temperature

How are these parameters measured (i.e., thermocouples, heat flux gauges, plate thermometers, etc.) and recorded (i.e., computer, strip charts, etc.)?

Thermocouples are used for measurements and recorded on computer.

Can your furnace(s) run positive, negative or a combination of both with respect to pressure inside the furnace? Do you have any limitations with respect to the pressure or the position of the neutral plane?

A combination of both. No limitation to the neutral plane.

Do you measure and record the furnace pressure?

Yes

Can you apply a load to the test sample?

No

If so, can this be done on each furnace?

Not applicable

Provide a brief description of the load mechanism (i.e., compression, tension, shear, axial, load limits, use load beam w/ jacks, etc.) for each furnace and/or type of building element.

Not applicable

How is the load carrying capability monitored during the test?

Not applicable

PART III – HORIZONTAL STRUCTURAL BUILDING ELEMENTS

Can you test for fire resistance of horizontal building elements such as floors, ceilings, beams, trusses, etc.?

No

Can you perform standard fire tests and nonstandard fire resistance tests on these elements?

How many furnaces for horizontal element tests do you have and what elements can be tested in each?

Besides testing discrete building elements such as floors or beams by themselves, can your furnace(s) accommodate combinations of elements such as floor-wall combination, column-beam connections, etc.?

What are the largest dimensions of test samples that you can test in each furnace for each type or combination of building element?

Do you have any restrictions with respect to what type of materials can be tested (i.e., wood, steel, concrete, composites, aluminum, fire-proofing materials, etc.)?

What type for fire exposure can your furnace(s) generate (i.e., ISO 834, ASTM E119, Hydrocarbon pool fire simulations, others)?

Do you have any limitations on the fire exposure?

Do you control the furnace exposure via temperatures and/or heat flux?

How are these parameters measured (i.e., thermocouples, heat flux gauges, plate thermometers, etc.) and recorded (i.e., computer, strip charts, etc.)?

Can your furnace(s) run positive or negative with respect to pressure inside the furnace? Do you have any limitations with respect to the pressure or the position of the neutral plane?

Do you measure and record the furnace pressure?

Can you apply a load to the test sample?

If so, can this be done on each furnace?

Provide a brief description of the load mechanism (i.e., compression, tension, shear, axial, load limits, dead load, live load, use loading jacks/frame, water containers, etc.) for each furnace and/or type of building element.

How is the load carrying capability monitored during the test?

PART IV – ADDITIONAL INFORMATION

Is your laboratory capable of and willing to performing tests for outside parties or do you test only for yourself or your industry?

Yes, we do conduct tests for overseas clients beside the local industries.

Do you have any restrictions on doing testing for outside parties?

No

Are there any specific contractual concerns with respect to performing testing for outside parties?

No

What is the typical turn-around time for conducting tests? We realize that cure times, construction, etc. may affect this but in general what is the timing.

4 to 6 weeks

Could you please provide a range of costs for some typical tests such as walls, columns, or floors? We realize many factors will influence this, but we are only looking for some general cost information.

Test fees of between US$2300 to US$3300.

Besides normal safety and environmental regulations, does your laboratory have any special regulations or rules concerning such items as testing, materials or disposal? If so, briefly details these.

We are expected to comply with the Ministry of Manpower and Ministry of Environment regulations, beside the laboratory acreditation under ISO/IEC 17025.

Besides, the furnaces described above, does your laboratory have the capability to perform full-scale, nonstandard, fire resistance tests of structures such as rooms or small buildings?

No.

Any other information that may assist us in our survey.

Our fire test laboratory is acredited by Singapore Accreditation Council (SAC) under the Singapore Laboratory Accreditation Scheme (SINGLAS). SAC is an independent organisation.

UL

PART I – GENERAL

Name of Laboratory

UNDERWRITERS LABORATORIES, INC

Location – City, State, Country

333 Pfingsten Road
Northbrook, IL 60062

Type of Organization – Government, Private, Industry, Non-profit, etc.

Not for profit

Contact information – Person, address, E-mail address, phone number, and fax number.

Robert M. Berhinig
333 Pfingsten Road
Northbrook, IL 60062

robert.m.berhinig@us.ul.com

Phone: +1 847-664-2292

Fax: +1-847-509-6392

Can the laboratory perform fire resistance testing of structural building elements? These include walls, columns, floors, ceilings, beams, trusses, connections etc.

Yes

If no – so state, complete Part I and return the form to Hughes Associates, Inc. as shown in Section V at the bottom of this Questionnaire.

If yes – please continue and address the parts that are applicable to your laboratory.

PART II – VERTICAL STRUCTURAL BUILDING ELEMENTS

Can you test for fire resistance of vertical building elements such as walls/columns, etc.?

Yes

Can you perform standard fire tests and nonstandard fire resistance tests on these elements?

Yes. There are temperature and pressure limits associated with each piece of test equipment. These conditions would need to be defined for the nonstandard tests.

How many furnaces for vertical element tests do you have and what elements can be tested in each?

Three

Column furnace: Inside dimensions of approximately 8 feet by 10 feet wide by 10 feet high.

Vertical Panel Furnace: Inside dimensions of sample test frame is approximately 15 feet wide by 10 feet high. This furnace is to test load and nonload bearing walls and fire door assemblies.

Vertical Panel Furnace: Inside dimensions of sample test frame is approximately 4.5 feet wide by 5.5 feet high.

Besides testing discrete building elements such as walls or columns by themselves, can your furnace(s) accommodate combinations of elements such as wall-floor combinations, column-beam connections, etc.?

The combinations of wall-floors and column beams may better be handled within our floor furnace. It is planned to add loading equipment to the column furnace during 2003. Presently only unloaded specimens can be tested in the column furnace.

What are the largest dimensions of test samples that you can test in each furnace for each type or combination of building element?

Column furnace – Approximately 9 feet long.

Vertical Panel Furnace – Approximately 14 feet wide by 10 feet high

Do you have any restrictions with respect to what type of materials can be tested (i.e., wood, steel, concrete, composites, aluminum, fire-proofing materials, etc.)?

No. All our furnaces are connected to smoke abatement equipment.

What type for fire exposure can your furnace(s) generate (i.e., ISO 834, ASTM E119, Hydrocarbon pool fire simulations, others)?

The large panel furnace is limited to ISO 834 / ASTM E119 exposures or lower temperature and pressure levels.

The column furnace is frequently used for hydrocarbon pool fire simulations. It was used to develop the requirements for Standard UL 1709.

Do you have any limitations on the fire exposure?

None other than the upper limits of the fuel source which excludes the large panel furnace from the rate of temperature rise required by the hydrocarbon pool fire simulation.

Do you control the furnace exposure via temperatures and/or heat flux?

Furnaces are controlled by temperatures because of test standard requirements. Heat flux measurements could be made.

How are these parameters measured (i.e., thermocouples, heat flux gauges, plate thermometers, etc.) and recorded (i.e., computer, strip charts, etc.)?

Temperatures are measured by thermocouples and recorded in a centralized data logging system. The data logging system is capable of handling 300 channels and is updated at a frequency of 1 second / iteration. Final processing of data is typically in Excel spreadsheet format.

Can your furnace(s) run positive, negative or a combination of both with respect to pressure inside the furnace? Do you have any limitations with respect to the pressure or the position of the neutral plane?

The furnaces can comply with the pressure requirements of ISO 834.

Do you measure and record the furnace pressure?

Yes

Can you apply a load to the test sample?

Yes

If so, can this be done on each furnace?

Samples on the large panel furnace can be loaded. These samples are usually 10 feet by 10 feet. Samples on the smaller vertical panel furnace can not be loaded.

Presently, samples in the column furnace cannot be loaded. It is anticipated loading equipment will installed in the column furnace during 2003.

Provide a brief description of the load mechanism (i.e., compression, tension, shear, axial, load limits, use load beam w/ jacks, etc.) for each furnace and/or type of building element.

Load is applied to wall by a series of hydraulic jacks.

How is the load carrying capability monitored during the test?

The load carrying capability of the sample is monitored by visual observations and by deflection measurements.

PART III – HORIZONTAL STRUCTURAL BUILDING ELEMENTS

Can you test for fire resistance of horizontal building elements such as floors, ceilings, beams, trusses, etc.?

Yes

Can you perform standard fire tests and nonstandard fire resistance tests on these elements?

Nonstandard tests can be conducted but there are limits to the capabilities of the furnace.

How many furnaces for horizontal element tests do you have and what elements can be tested in each?

Three.

One furnace can accommodate samples approximately 14 feet wide by 17.5 feet long. The depth of the furnace can vary but the most common depth is approximately 6 feet.

The other two furnaces are smaller with sample sizes approximately 3 feet by 3 feet. Loads cannot be applied on these smaller samples.

Besides testing discrete building elements such as floors or beams by themselves, can your furnace(s) accommodate combinations of elements such as floor-wall combination, column-beam connections, etc.?

Yes, although this would not be a typical test sample. We have tested floor wall combinations in our large horizontal furnace.

What are the largest dimensions of test samples that you can test in each furnace for each type or combination of building element?

The largest floor samples are approximately 14 feet by 17 feet. The longest beam samples are approximately 17 feet.

Do you have any restrictions with respect to what type of materials can be tested (i.e., wood, steel, concrete, composites, aluminum, fire-proofing materials, etc.)?

None. All furnaces are connected to smoke abatement equipment.

What type for fire exposure can your furnace(s) generate (i.e., ISO 834, ASTM E119, Hydrocarbon pool fire simulations, others)?

ISO 834 and ASTM E119. It is planned to expand the capabilities of the large horizontal furnace to be able to conduct hydrocarbon pool fire simulations during 2003.

Do you have any limitations on the fire exposure?

The limitations are controlled by the fuel supply. It is anticipated that by mid-2003, the floor furnace will be able to provide the temperature rate of rise required by the hydrocarbon pool fire simulation. The furnaces can provide any fire exposure that requires a time–temperature less than that required by ISO 834 or ASTM E119.

Do you control the furnace exposure via temperatures and/or heat flux?

Furnaces are controlled by temperatures because of test standard requirements Heat flux measurements could be made.

How are these parameters measured (i.e., thermocouples, heat flux gauges, plate thermometers, etc.) and recorded (i.e., computer, strip charts, etc.)?

Temperatures are measured by thermocouples and recorded in a centralized data logging system. The data logging system is capable of handling 300 channels and is updated at a frequency of 1 second / iteration. Final processing of data is typically in Excel spreadsheet format.

Can your furnace(s) run positive or negative with respect to pressure inside the furnace? Do you have any limitations with respect to the pressure or the position of the neutral plane?

The large furnace can comply with the requirements of ISO 834. The neutral pressure plane can be located close to exposed surface of the test specimen.

Do you measure and record the furnace pressure?

Yes.

Can you apply a load to the test sample?

Yes

If so, can this be done on each furnace?

Only the large horizontal furnace.

Provide a brief description of the load mechanism (i.e., compression, tension, shear, axial, load limits, dead load, live load, use loading jacks/frame, water containers, etc.) for each furnace and/or type of building element.

A combination of methods are used depending the requirements of the sample being tested. Equipment varies from concrete blocks, water containers and hydraulic jacks.

How is the load carrying capability monitored during the test?

For the hydraulic system, the pressure is monitored. When using water containers or other somewhat static means monitoring is by visual observations.

PART IV – ADDITIONAL INFORMATION

Is your laboratory capable of and willing to performing tests for outside parties or do you test only for yourself or your industry?

We are willing to conduct tests for outside parties.

Do you have any restrictions on doing testing for outside parties?

No. UL has no general restrictions on doing testing for outside parties (other than those foreign entities with whom U.S. corporations are barred from doing business by government regulations).

Are there any specific contractual concerns with respect to performing testing for outside parties?

No. UL typically employs its standard form contracts to support its testing services.

What is the typical turn-around time for conducting tests? We realize that cure times, construction, etc. may affect this but in general what is the timing.

Within two weeks upon receiving the materials necessary to construct the sample. There is no delay in testing the sample when the sample has reached required curing.

Could you please provide a range of costs for some typical tests such as walls, columns, or floors? We realize many factors will influence this, but we are only looking for some general cost information.

Concrete floors with protected steel - $48,000.00

Walls - $22,000.00

Columns - $9,000.00

These costs assume UL staff will construct the assemblies except for the application of spray applied fire resistive coatings which would be by an outside contractor. The contractor's fee is included in the cost estimates.

Besides normal safety and environmental regulations, does your laboratory have any special regulations or rules concerning such items as testing, materials or disposal? If so, briefly details these.

UL follows MSDS regulations.

Besides, the furnaces described above, does your laboratory have the capability to perform full-scale, nonstandard, fire resistance tests of structures such as rooms or small buildings?

Yes. UL a several rooms dedicated to fire testing. The most versatile is 120 feet long by 120 feet wide with a ceiling height that can vary from 5 feet to 48 feet.

Any other information that may assist us in our survey.

Please see the attached folder. UL's fire testing services go well beyond the typical test equipment associated with testing fire resistive building assemblies. The enclosed "fact sheets" highlight several additional areas such as performance of sprinkler systems and their components, flammability of materials, extinguishing systems and large-scale fire research.

USG

PART I – GENERAL

Name of Laboratory USG Research & Technology Center

Location – City, State, Country Libertyville, IL USA 60048

Type of Organization – Government, Private, Industry, Non-profit, etc. Private (Corporate Research Facility)

Contact information – Person, address, E-mail address, phone number, and fax number. Rich Kaczkowski, 700 N. Highway 45, Libertyville, IL 60048, PH 847-970-5255, FAX 847-970-5299

Can the laboratory perform fire resistance testing of structural building elements? These include walls, columns, floors, ceilings, beams, trusses, connections etc. Yes – as described below

If no – so state, complete Part I and return the form to Hughes Associates, Inc. as shown in Section V at the bottom of this Questionnaire.

If yes – please continue and address the parts that are applicable to your laboratory.

PART II – VERTICAL STRUCTURAL BUILDING ELEMENTS

Can you test for fire resistance of vertical building elements such as walls/columns, etc.? Yes – both load bearing and non load bearing walls can be tested

Can you perform standard fire tests and nonstandard fire resistance tests on these elements? Yes – our wall furnace complies with ASTM E119

How many furnaces for vertical element tests do you have and what elements can be tested in each? One vertical furnace for wall testing

Besides testing discrete building elements such as walls or columns by themselves, can your furnace(s) accommodate combinations of elements such as wall-floor combinations, column-beam connections, etc.? No

What are the largest dimensions of test samples that you can test in each furnace for each type or combination of building element? Wall size is limited to 10' x 10'

Do you have any restrictions with respect to what type of materials can be tested (i.e., wood, steel, concrete, composites, aluminum, fire-proofing materials, etc.)? No. The only limitations are safety and practicality.

What type for fire exposure can your furnace(s) generate (i.e., ISO 834, ASTM E119, Hydrocarbon pool fire simulations, others)? Any time-temperature curve can be programmed, but max temp is limited to 2200 deg F.

Do you have any limitations on the fire exposure? See above

Do you control the furnace exposure via temperatures and/or heat flux? Furnace exposure is temperature controlled.

How are these parameters measured (i.e., thermocouples, heat flux gauges, plate thermometers, etc.) and recorded (i.e., computer, strip charts, etc.)? Measured via thermocouples, recorded via computerized data acquisition system.

Can your furnace(s) run positive, negative or a combination of both with respect to pressure inside the furnace? Do you have any limitations with respect to the pressure or the position of the neutral plane? Yes. Neutral plane can vary from lower $1/3^{rd}$ point to slightly above top of test assembly.

Do you measure and record the furnace pressure? Yes

Can you apply a load to the test sample? Yes

If so, can this be done on each furnace? N/A. We only have one wall furnace.

Provide a brief description of the load mechanism (i.e., compression, tension, shear, axial, load limits, use load beam w/ jacks, etc.) for each furnace and/or type of building element. Load beam with mechanical jacks.

How is the load carrying capability monitored during the test?

PART III – HORIZONTAL STRUCTURAL BUILDING ELEMENTS

Can you test for fire resistance of horizontal building elements such as floors, ceilings, beams, trusses, etc.? Yes – but size is limited and horizontal furnace does not comply with ASTM E119 size requirements.

Can you perform standard fire tests and nonstandard fire resistance tests on these elements? Furnace size does not meet ASTM E119 requirements.

How many furnaces for horizontal element tests do you have and what elements can be tested in each? One. It has been used to test floor-ceiling and roof-ceiling assemblies.

Besides testing discrete building elements such as floors or beams by themselves, can your furnace(s) accommodate combinations of elements such as floor-wall combination, column-beam connections, etc.? No.

What are the largest dimensions of test samples that you can test in each furnace for each type or combination of building element? Approximately 7'-6" by 8'-6"

Do you have any restrictions with respect to what type of materials can be tested (i.e., wood, steel, concrete, composites, aluminum, fire-proofing materials, etc.)? No. The only limitations are safety and practicality

What type for fire exposure can your furnace(s) generate (i.e., ISO 834, ASTM E119, Hydrocarbon pool fire simulations, others)? Any time-temperature curve can be programmed, but max temp is limited to 2200 deg F.

Do you have any limitations on the fire exposure? See above

Do you control the furnace exposure via temperatures and/or heat flux? Furnace exposure is temperature controlled. How are these parameters measured (i.e., thermocouples, heat flux gauges, plate thermometers, etc.) and recorded (i.e., computer, strip charts, etc.)? Measured via thermocouples, recorded via computerized data acquisition system.

Can your furnace(s) run positive or negative with respect to pressure inside the furnace? Do you have any limitations with respect to the pressure or the position of the neutral plane? Yes. Neutral plane can vary from lower $1/3^{rd}$ point to slightly above top of test assembly.

Do you measure and record the furnace pressure? Yes

Can you apply a load to the test sample? Yes

If so, can this be done on each furnace? N/A. We only have one wall furnace

Provide a brief description of the load mechanism (i.e., compression, tension, shear, axial, load limits, dead load, live load, use loading jacks/frame, water containers, etc.) for each furnace and/or type of building element. Dead load is applied via water containers

How is the load carrying capability monitored during the test? Since load is applied as dead weight, no monitoring is required.

PART IV – ADDITIONAL INFORMATION

Is your laboratory capable of and willing to performing tests for outside parties or do you test only for yourself or your industry? We are a corporate Research facility and do testing to support USG Corporation. We typically do not do testing for hire.

Do you have any restrictions on doing testing for outside parties? See above

Are there any specific contractual concerns with respect to performing testing for outside parties? See above

What is the typical turn-around time for conducting tests? We realize that cure times, construction, etc. may affect this but in general what is the timing. N/A

Could you please provide a range of costs for some typical tests such as walls, columns, or floors? We realize many factors will influence this, but we are only looking for some general cost information. N/A

Besides normal safety and environmental regulations, does your laboratory have any special regulations or rules concerning such items as testing, materials or disposal? If so, briefly details these. N/A

Besides, the furnaces described above, does your laboratory have the capability to perform full-scale, nonstandard, fire resistance tests of structures such as rooms or small buildings? No.

Any other information that may assist us in our survey.

What are the largest dimensions of test samples that you can test in each furnace for each type or combination of building element? Approximately 7'-6" by 8'-6"

Do you have any restrictions with respect to what type of materials can be tested (i.e., wood, steel, concrete, composites, aluminum, fire-proofing materials, etc.)? No. The only limitations are safety and practicality

What type for fire exposure can your furnace(s) generate (i.e., ISO 834, ASTM E119, Hydrocarbon pool fire simulations, others)? Any time-temperature curve can be programmed, but max temp is limited to 2200 deg F.

Do you have any limitations on the fire exposure? See above

Do you control the furnace exposure via temperatures and/or heat flux? Furnace exposure is temperature controlled. How are these parameters measured (i.e., thermocouples, heat flux gauges, plate thermometers, etc.) and recorded (i.e., computer, strip charts, etc.)? Measured via thermocouples, recorded via computerized data acquisition system.

Can your furnace(s) run positive or negative with respect to pressure inside the furnace? Do you have any limitations with respect to the pressure or the position of the neutral plane? Yes. Neutral plane can vary from lower $1/3^{rd}$ point to slightly above top of test assembly.

Do you measure and record the furnace pressure? Yes

Can you apply a load to the test sample? Yes

If so, can this be done on each furnace? N/A. We only have one wall furnace

Provide a brief description of the load mechanism (i.e., compression, tension, shear, axial, load limits, dead load, live load, use loading jacks/frame, water containers, etc.) for each furnace and/or type of building element. Dead load is applied via water containers

How is the load carrying capability monitored during the test? Since load is applied as dead weight, no monitoring is required.

PART IV – ADDITIONAL INFORMATION

Is your laboratory capable of and willing to performing tests for outside parties or do you test only for yourself or your industry? We are a corporate Research facility and do testing to support USG Corporation. We typically do not do testing for hire.

Do you have any restrictions on doing testing for outside parties? See above

Are there any specific contractual concerns with respect to performing testing for outside parties? See above

What is the typical turn-around time for conducting tests? We realize that cure times, construction, etc. may affect this but in general what is the timing. N/A

Could you please provide a range of costs for some typical tests such as walls, columns, or floors? We realize many factors will influence this, but we are only looking for some general cost information. N/A

Besides normal safety and environmental regulations, does your laboratory have any special regulations or rules concerning such items as testing, materials or disposal? If so, briefly details these. N/A

Besides, the furnaces described above, does your laboratory have the capability to perform full-scale, nonstandard, fire resistance tests of structures such as rooms or small buildings? No.

Any other information that may assist us in our survey.

www.ingramcontent.com/pod-product-compliance
Lightning Source LLC
Chambersburg PA
CBHW081829170526

45167CB00007B/2762